"Never shy about tackling big questions, veteran evolutionary biologist Wilson (*The Creation: An Appeal to Save Life on Earth*) delivers his thoughtful if contentious explanation of why humans rule the Earth. . . . Wilson succeeds in explaining his complex ideas, so attentive readers will receive a deeply satisfying exposure to a major scientific controversy." —*Kirkus Reviews*, starred review

"*The Social Conquest of Earth* has set off a scientific furor. . . . The controversy is fueled by a larger debate about the evolution of altruism. Can true altruism even exist? Is generosity a sustainable trait? Or are living things inherently selfish, our kindness nothing but a mask? This is science with existential stakes." —Jonah Lehrer, *The New Yorker*

"With bracing insights into instinct, language, organized religion, the humanities, science, and social intelligence, this is a deeply felt, powerfully written, and resounding inquiry into the human condition." —*Booklist*, starred review

"That Wilson provides nimble, lucid responses to the three core questions speaks volumes about his intellectual rigor. That he covers all of this heady terrain in less than 300 pages of text speaks volumes about his literary skill." —Larry Lebowitz, *Miami Herald*

"Wilson frames *The Social Conquest of Earth* as a dialogue with painter Paul Gauguin, who penned on the canvas of his 1897 Tahitian masterpiece: 'Where do we come from? What are we? Where are we going?' . . . Wilson attempts to answer Gauguin . . . by embracing the existential questioning of the humanities without sacrificing the 'unrelenting application of reason' at the core of empirical science." —Alyssa A. Botelho, *Harvard Crimson*

"The Harvard University naturalist and Pulitzer Prize winner angered many colleagues two years ago, when he repudiated a concept within evolutionary theory that he had brought to prominence. Known as kin selection or inclusive fitness, the half-century-old idea helped to explain the puzzling existence of altruism among animals. Why, for instance, do some birds help their parents raise chicks instead of having chicks of their own? Why are worker ants sterile? The answer, according to kin selection theory, has been that aiding your relatives can sometimes spread your common genes faster than bearing offspring of your own. In *The Social Conquest of Earth*, Wilson offers a full explanation of his latest thinking on evolution. Group dynamics, not selfish genes, drive altruism, he argues: 'Colonies of cheaters lose to colonies of cooperators.' As the cooperative colonies dominate and multiply, so do their alleged 'altruism' genes. Wilson uses what he calls 'multilevel selection'—group and individual selection combined—to discuss the emergence of the creative arts and humanities, morality, religion, language and the very nature of humans. Along the way, he pauses to reject religion, decry the way humans have despoiled the environment and, in something of a non sequitur, dismiss the need for manned space exploration. The book is bound to stir controversy on these and other subjects for years to come." —Sandra Upson and Anna Kuchment, *Scientific American*

"Pretty much anything Wilson writes is well worth reading, and his latest, *The Social Conquest of Earth*, is no exception. . . . Read the master biologist himself in this marvelous book."—Michael Shermer, *The Daily*

"With his probing curiosity, his dazzling research, his elegant prose and his deep commitment to biodiversity, Pulitzer Prize–winning biologist (*The Ants*) and novelist (*Anthill*) Edward O. Wilson has spent his life searching for the evolutionary paths by which humans developed and passed along the social behaviors that best promote the survival of our species. His eloquent, magisterial and compelling new book offers a kind of summing-up of his magnificent career. . . . While not everyone will agree with Wilson's provocative and challenging conclusions, everyone who engages with his ideas will discover sparkling gems of wisdom uncovered by the man who is our Darwin and our Thoreau." —Henry L. Carrigan Jr., *BookPage*

TH[illegible] EARTH

"Religion. Sports. War. Biologist E. O. Wilson says our drive to join a group—and to fight for it—is what makes us human." —*Newsweek*

"Wilson has done an impressive job of pulling all this evidence together and analyzing it. His interdisciplinary approach, his established scholarship, and his willingness to engage hot-button issues are all much in evidence in *The Social Conquest of Earth*. . . . His reflections on this subject are varied, original, and thought provoking—as is the rest of his book." —Carl Coon, *The Humanist*

"Wilson's examples of insect eusociality are dazzling. . . . There are obvious parallels with human practices like war and agriculture, but Wilson is also sensitive to the differences. . . . This book offers a detailed reconstruction of what we know about the evolutionary histories of these two very different conquerors. Wilson's careful and clear analysis reminds us that scientific accounts of our origins aren't just more accurate than religious stories; they are also a lot more interesting." —Paul Bloom, *New York Times Book Review*

"Edward O. Wilson's passionate curiosity—the hallmark of his remarkable career—has led him to these urgent reflections on the human condition. At the core of *The Social Conquest of Earth* is the unresolved, unresolvable tension in our species between selfishness and altruism. Wilson brilliantly analyzes the force, at once creative and destructive, of our biological inheritance and daringly advances a grand theory of the origins of human culture. This is a wonderful book for anyone interested in the intersection of science and the humanities." —Stephen Greenblatt, author of *The Swerve: How the World Became Modern*

"A sweeping argument about the biological origins of complex human culture. It is full of both virtuosity and raw, abrupt assertions that are nonetheless well-crafted and captivating. . . . It is fascinating to see such a distinguished scientist optimistic about the future."
—Michael Gazzaniga, *Wall Street Journal*

"Once again, Edward O. Wilson has written a book combining the qualities that have brought his previous books Pulitzer Prizes and millions of readers: a big but simple question, powerful explanations, magisterial knowledge of the sciences and humanities, and beautiful writing understandable to a wide public." —Jared Diamond, Pulitzer Prize–winning author of *Guns, Germs, and Steel*

"Wilson's newest theory . . . could transform our understanding of human nature—and provide hope for our stewardship of the planet. . . . His new book . . . is not limited to the discussion of evolutionary biology, but ranges provocatively through the humanities. . . . Its impact on the social sciences could be as great as its importance for biology, advancing human self-understanding in ways typically associated with the great philosophers." —Howard W. French, *The Atlantic*

"A monumental exploration of the biological origins of the human condition."
—James D. Watson

"*The Social Conquest of Earth* is a huge, deep, thrilling work, presenting a radically new but cautiously hopeful view of human evolution, human nature, and human society. No one but Edward O. Wilson could bring together such a brilliant synthesis of biology and the humanities to shed light on the origins of language, religion, art, and all of human culture." —Oliver Sacks, author of *The Mind's Eye*

"Biologist E. O. Wilson's brilliant new volume, *The Social Conquest of Earth,* could more aptly be entitled 'Biology's Conquest of Science.' Drawing on his deep understanding of entomology and his extraordinarily broad knowledge of the natural and social sciences, Wilson makes a strong case for the synthesis of knowledge across disciplines. Understanding the biological origin of what makes us human can help us to build better theories of social and psychological interaction; in turn, understanding how other social species have evolved may help us to better understand the origin of our own. But the main reason that Wilson's book is successful is that he also brings into biology the best of what social science has to offer."

—James H. Fowler, *Nature* magazine

"An ambitious and thoroughly engaging work that's certain to generate controversy within the walls of academia and without. . . . Provocative, eloquent and unflinchingly forthright, Wilson remains true to form, producing a book that's anything but dull and bound to receive plenty of attention from supporters and critics alike."

—Colin Woodard, *Washington Post*

" 'Where do we come from? What are we? Where are we going?' Those famous questions, inscribed by Paul Gauguin in his giant Tahitian painting of 1897, introduce *The Social Conquest of Earth.* Their choice proclaims Edward O. Wilson's ambitions for his splendid book, in which he sums up 60 distinguished years of research into the evolution of human beings and social insects." —Clive Cookson, *Financial Times*

"Wilson is a brilliant stylist, and his account of the rise of *Homo sapiens* and our species' conquest of Earth is informative, thrilling, and utterly captivating." —Rudy M. Baum, *Chemical & Engineering News*

"What Wilson ends up doing is so profound that the last eight chapters . . . could stand alone as a separate book, because what he ends up doing is no less than defining human nature itself."

—Robert M. Knight, *Washington Independent Review of Books*

"Reading E. O. Wilson's *The Social Conquest of Earth* is a revolutionary look at who we are, where we've come from and where we're going. It's very hopeful in that he suggests that we have the capacity to learn to live within the planet's means. I personally call this the sweet spot in history. Never before have we had the knowledge and opportunity as good as we have now to make change. The great message Wilson conveys is that there's still time."
—Kate Murphy,
New York Times Sunday Review

"I just finished *The Social Conquest of Earth*, a fabulous book."
—President Bill Clinton, *New York Times*

The Social Conquest *of* Earth

Also by Edward O. Wilson

Genesis: The Deep Origin of Societies (2019)

The Origins of Creativity (2017)

Half-Earth: Our Planet's Fight for Life (2016)

The Meaning of Human Existence (2014)

Letters to a Young Scientist (2013)

The Leafcutter Ants: Civilization by Instinct, with Bert Hölldobler (2011)

Kingdom of Ants: José Celestino Mutis and the Dawn of Natural History in the New World, with José M. Gómez Durán (2010)

Anthill: A Novel (2010)

The Superorganism: The Beauty, Elegance, and Strangeness of Insect Societies, with Bert Hölldobler (2009)

The Creation: An Appeal to Save Life on Earth (2006)

Nature Revealed: Selected Writings, 1949–2006 (2006)

From So Simple a Beginning: The Four Great Books of Darwin, edited with introductions (2005)

Pheidole *in the New World: A Hyperdiverse Ant Genus* (2003)

The Future of Life (2002)

Biological Diversity: The Oldest Human Heritage (1999)

Consilience: The Unity of Knowledge (1998)

In Search of Nature (1996)

Journey to the Ants: A Story of Scientific Exploration, with Bert Hölldobler (1994)

Naturalist (1994); new edition, 2006

The Diversity of Life (1992)

The Ants, with Bert Hölldobler (1990); Pulitzer Prize, General Nonfiction, 1991

Success and Dominance in Ecosystems: The Case of the Social Insects (1990)

Biophilia (1984)

Promethean Fire: Reflections on the Origin of the Mind, with Charles J. Lumsden (1983)

Genes, Mind, and Culture: The Coevolutionary Process, with Charles J. Lumsden (1981)

On Human Nature (1978); Pulitzer Prize, General Nonfiction, 1979

Caste and Ecology in the Social Insects, with George F. Oster (1978)

Sociobiology: The New Synthesis (1975); new edition, 2000

A Primer of Population Biology, with William H. Bossert (1971)

The Insect Societies (1971)

The Theory of Island Biogeography, with Robert MacArthur (1967)

The Social Conquest *of* Earth

Edward O. Wilson

Liveright Publishing Corporation
A Division of W. W. Norton & Company
New York · London

Printed in the United States of America
First published as a Liveright paperback 2013

For information about permission to reproduce selections from this book, write to Permissions, Liveright Publishing Corporation, a division of W. W. Norton & Company, Inc., 500 Fifth Avenue, New York, NY 10110

For information about special discounts for bulk purchases, please contact W. W. Norton Special Sales at specialsales@wwnorton.com or 800-233-4830

Manufacturing by LSC Harrisonburg
Book design by Judith Stagnitto Abbate / Abbate Design
Production manager: Anna Oler

Library of Congress Cataloging-in-Publication Data

Wilson, Edward O.
The social conquest of earth / Edward O. Wilson. — 1st ed.
p. cm.
Includes bibliographical references and index.
ISBN 978-0-87140-413-8 (hardcover)
1. Social evolution—Philosophy. 2. Human evolution—Philosophy.
3. Evolution (Biology)—Philosophy. I. Title.
GN360.W53 2012
599.93'8—dc23

2011052680

ISBN 978-0-87140-363-6 pbk.

Liveright Publishing Corporation
500 Fifth Avenue, New York, N.Y. 10110
www.wwnorton.com

W. W. Norton & Company Ltd.
15 Carlisle Street, London W1D 3BS

5 6 7 8 9 0

Contents

Prologue 1

I: Why Does Advanced Social Life Exist?

1. The Human Condition 7

II: Where Do We Come From?

2. The Two Paths to Conquest 13
3. The Approach 21
4. The Arrival 33
5. Threading the Evolutionary Maze 45
6. The Creative Forces 49
7. Tribalism Is a Fundamental Human Trait 57
8. War as Humanity's Hereditary Curse 62
9. The Breakout 77
10. The Creative Explosion 85
11. The Sprint to Civilization 97

III: How Social Insects Conquered the Invertebrate World

12. The Invention of Eusociality 109
13. Inventions That Advanced the Social Insects 120

IV: The Forces of Social Evolution

14. The Scientific Dilemma of Rarity 133
15. Insect Altruism and Eusociality Explained 139
16. Insects Take the Giant Leap 148
17. How Natural Selection Creates Social Instincts 158
18. The Forces of Social Evolution 166
19. The Emergence of a New Theory of Eusociality 183

V: What Are We?

20. What Is Human Nature? 191
21. How Culture Evolved 212
22. The Origins of Language 225
23. The Evolution of Cultural Variation 236
24. The Origins of Morality and Honor 241
25. The Origins of Religion 255
26. The Origins of the Creative Arts 268

VI: Where Are We Going?

27. A New Enlightenment 287

Acknowledgments 301

References 303

Index 327

The Social Conquest *of* Earth

Prologue

THERE IS NO GRAIL more elusive or precious in the life of the mind than the key to understanding the human condition. It has always been the custom of those who seek it to explore the labyrinth of myth: for religion, the myths of creation and the dreams of prophets; for philosophers, the insights of introspection and reasoning based upon them; for the creative arts, statements based upon a play of the senses.

Great visual art in particular is the expression of a person's journey, an evocation of feeling that cannot be put into words. Perhaps in the hitherto hidden lies deeper, more essential meaning. Paul Gauguin, hunter of secrets and famed Maker of Myths (as he has been called), made this attempt. His story is a worthy backdrop for the modern answer to be offered in the present work.

Late in 1897, at Punaauia, three miles from the Tahitian port of Papeete, Gauguin sat down to put on canvas his largest and most important painting. He was weak from syphilis and a series of debilitating heart attacks. His funds were nearly gone, and he was depressed by the news that his daughter Aline had recently died of pneumonia in France.

Gauguin knew his time was running out. He meant this painting to be his last. And so when he finished, he went into the mountains behind Papeete to commit suicide. He carried with him a vial of arsenic he had stored, perhaps unaware of how painful death by this poison can be. He intended to hide himself before he took it, so that his corpse would not be found right away and instead would be eaten by ants.

But then he relented, and returned to Punaauia. Although there was very little left to his life, he had decided to soldier on. To survive, he took a six-franc-a-day job in Papeete as a clerk in the Office of Public Works and Surveys. In 1901, he sought even greater isolation, moving to the little island of Hiva Oa in the faraway Marquesas archipelago. Two years later, while embroiled in legal problems, Paul Gauguin died of syphilitic heart failure. He was buried in the Catholic cemetery on Hiva Oa.

"I am a savage," he wrote a magistrate a few days before the end. "And civilized people suspect this, for in my works there is nothing so surprising and baffling as this 'savage in spite of myself' aspect."

Gauguin had come to French Polynesia, to this almost impossible end of the world (only Pitcairn and Easter Island are more remote), to find both peace and a new frontier of artistic expression. He attained the second, if not the first.

Gauguin's journey of body and mind was unique among major artists of his era. Born in Paris in 1848, he was raised in Lima and then Orléans by his half-Peruvian mother. This ethnic mix gave a hint of what was to come. As a young man he joined the French merchant marines and traveled around the world for six years. During this period, in 1870–71, he saw action in the Franco-Prussian War, in the Mediterranean and North Sea. Back in Paris he at first gave little thought to art, instead becoming a stockbroker under the guidance of his wealthy guardian Gustave Arosa. His interest in art was sparked and sustained by Arosa, a major collector of French art, including the latest works of impressionism. When the French stock market crashed in January 1882 and his own bank failed, Gauguin turned to painting and began to develop his considerable talent. Nurtured in impressionism by painters of undoubted greatness—Pissarro, Cézanne, Van Gogh, Manet, Seurat, Degas—he strove to join their ranks. As he traveled about, from Pontoise to Rouen, from Pont-Aven to Paris, he created portraits, still lifes, landscapes, in work increasingly phantasmagoric, portending the Gauguin who was to emerge.

But Gauguin was disappointed with the result, and lingered only a short time in the company of his dazzling contemporaries. He had not grown rich and famous with his own efforts, even though, as he later declared, he knew he was a great artist. He longed for a simpler, easier life to meet this destiny. Paris, he wrote in 1886, "is a wasteland for a poor man. . . . I am going to Panama to live the life of a native. . . . I shall take my paints and brushes and reinvigorate myself far from the company of men."

It was not just poverty that drove Gauguin from civilization. He was at heart a restless soul, an adventurer, ever anxious to find what lay beyond the place he lived. In art, he was accordingly an experimentalist. In his wanderings he was drawn to the exoticism of non-Western cultures, and wanted to immerse himself in them in search of new modes of visual expression. He spent time in Panama and then Martinique. Returning home, he applied for a position in the French-ruled province of Tonkin, now northern Vietnam. When that failed, he turned at last to French Polynesia, the ultimate paradise.

On June 9, 1891, Gauguin arrived at Papeete and immersed himself in the indigenous culture. In time he became an advocate of native rights, and therefore a troublemaker in the eyes of the colonial authorities. Of vastly greater importance, he pioneered the new style called primitivism: flat, pastoral, often violently colorful, simple and direct, and authentic.

We cannot escape the conclusion, however, that Gauguin sought more than just this new style. He was also deeply interested in the human condition, in what it truly is and how to portray it. The venues of metropolitan France, especially Paris, were a domain of a thousand voices shouting for attention, where intellectual and artistic life was ruled by recognized authorities, each rooted in his own small acreage of expertise. No one, he felt, could make a new unity out of that cacophony.

Such might be done, however, in the vastly simpler yet still wholly functional world of Tahiti. There one might possibly cut down to the bedrock of the human condition. In this respect Gauguin was one with Henry David Thoreau, who earlier had retreated to his tiny cabin on the edge of Walden Pond "to front only the essential facts of life, and see if I could not learn what it had to teach . . . to cut a broad swath and shave close, to drive life into a corner, and reduce it to its lowest terms."

That perception is best expressed by Gauguin on his twelve-foot-wide masterwork. Look closely at its details. It contains a row of figures arrayed in front of a faint mélange of Tahitian landscapes, mountain and sea. Most of the figures are female (this being the Tahitian Gauguin). Variously realistic and surreal, they represent the human life cycle. The artist intends for us to scan from right to left. A baby at the far right represents birth. An adult of ambiguous sex has been placed in the center, arms raised, a symbol of individual self-recognition. Nearby to the left a young couple picking and eating apples are the Adam-and-Eve archetype, in quest of knowledge. On the far left, representing death, an old woman is hunched in anguish and despair (thought to have been inspired by Albrecht Dürer's 1514 engraving *Melancholia*).

A blue-tinted idol stares at us from the left background, arms lifted ritualistically, perhaps benign, or perhaps malignant. Gauguin himself described its meaning with telling poetic ambiguity.

> The Idol is there not as a literary explanation, but as a statue, less statue perhaps than the animal figures; less animal too, becoming one in my dream, in front of my hut, with the whole of nature, dominating *our primitive soul*, the imaginary consolation of our sufferings and what they contain of the value and the uncomprehending before the mystery of our origins and our future. (Gauguin's emphasis)

On the upper left corner of the canvas he wrote the famous title, *D'où Venons Nous / Que Sommes Nous / Où Allons Nous.*

The painting is not an answer. It is a question.

I

Why Does Advanced Social Life Exist?

· 1 ·

The Human Condition

"W*here do we come from?*" "*What are we?*" "*Where are we going?*" Conceived in ultimate simplicity by Paul Gauguin on the canvas of his Tahitian masterpiece, these are in fact the central problems of religion and philosophy. Will we ever be able to solve them? Sometimes it seems not. Yet perhaps we can.

Humanity today is like a waking dreamer, caught between the fantasies of sleep and the chaos of the real world. The mind seeks but cannot find the precise place and hour. We have created a Star Wars civilization, with Stone Age emotions, medieval institutions, and god-like technology. We thrash about. We are terribly confused by the mere fact of our existence, and a danger to ourselves and to the rest of life.

Religion will never solve this great riddle. Since Paleolithic times each tribe—of which there have been countless thousands—invented its own creation myth. During this long dreamtime of our ancestors, supernatural beings spoke to shamans and prophets. They identified themselves to the mortals variously as God, a tribe of Gods, a divine family, the Great Spirit, the Sun, ghosts of the forebears, supreme serpents, hybrids of sundry animals, chimeras of men and beasts, omnipotent sky spiders—anything, everything that could be conjured by the dreams, hallucinogens, and fertile imaginations of the spiritual leaders. They were shaped in part by the environments of those who invented them. In Polynesia, gods pried the sky apart from the ground and sea, and the creation of life and humanity followed. In the desert-dwelling patriarchies of Judaism, Christianity, and

Islam, prophets conceived, not surprisingly, a divine, all-powerful patriarch who speaks to his people through sacred scripture.

The creation stories gave the members of each tribe an explanation for their existence. It made them feel loved and protected above all other tribes. In return, their gods demanded absolute belief and obedience. And rightly so. The creation myth was the essential bond that held the tribe together. It provided its believers with a unique identity, commanded their fidelity, strengthened order, vouchsafed law, encouraged valor and sacrifice, and offered meaning to the cycles of life and death. No tribe could long survive without the meaning of its existence defined by a creation story. The option was to weaken, dissolve, and die. In the early history of each tribe, the myth therefore had to be set in stone.

The creation myth is a Darwinian device for survival. Tribal conflict, where believers on the inside were pitted against infidels on the outside, was a principal driving force that shaped biological human nature. The truth of each myth lived in the heart, not in the rational mind. By itself, mythmaking could never discover the origin and meaning of humanity. But the reverse order is possible. The discovery of the origin and meaning of humanity might explain the origin and meaning of myths, hence the core of organized religion.

Can these two worldviews ever be reconciled? The answer, to put the matter honestly and simply, is no. They cannot be reconciled. Their opposition defines the difference between science and religion, between trust in empiricism and belief in the supernatural.

If the great riddle of the human condition cannot be solved by recourse to the mythic foundations of religion, neither will it be solved by introspection. Unaided rational inquiry has no way to conceive its own process. Most of the activities of the brain are not even perceived by the conscious mind. The brain is a citadel, as Darwin once put it, that cannot be taken by direct assault.

Thinking about thinking is the core process of the creative arts, but it tells us very little about *how* we think the way we do, and nothing of *why* the creative arts originated in the first place. Consciousness, having evolved over millions of years of life-and-death struggle, and moreover because of that struggle, was not designed for self-examination. It was designed for survival and reproduction. Conscious thought is driven by emotion; to the purpose of survival and reproduction, it is ultimately and wholly committed. The intricate distortions of the mind may be transmitted by the creative arts in fine detail, but they are constructed as though human nature never had an evolutionary history. Their powerful metaphors have brought us no closer to solving the riddle than did the dramas and literature of ancient Greece.

Scientists, scouting the perimeters of the citadel, search for potential breaches in its walls. Having broken through with technology designed for that purpose, they now read the codes and track the pathways of billions of nerve cells. Within a generation, we likely will have progressed enough to explain the physical basis of consciousness.

But—when the nature of consciousness is solved, will we then know what we are and where we came from? No, we will not. To understand the physical operations of the brain to their foundations brings us close to the grail. To find it, however, we need far more knowledge collected from both science and the humanities. We need to understand how the brain evolved the way it did, and why.

Moreover, we look in vain to philosophy for the answer to the great riddle. Despite its noble purpose and history, pure philosophy long ago abandoned the foundational questions about human existence. The query itself is a reputation killer. It has become a Gorgon for philosophers, upon whose visage even the best thinkers fear to gaze. They have good reason for their aversion. Most of the history of philosophy consists of failed models of the mind. The field of discourse is strewn with the wreckage of theories of consciousness. After the decline of logical positivism in the middle of the twentieth century, and the attempt of this movement to blend science and logic into a closed system, professional philosophers dispersed in an intellectual diaspora. They emigrated into the more tractable disciplines not yet colonized by science—intellectual history, semantics, logic, foundational mathematics, ethics, theology, and, most lucratively, problems of personal life adjustment.

Philosophers flourish in these various endeavors, but for the time being at least, and by a process of elimination, the solution of the riddle has been left to science. What science promises, and has already supplied in part, is the following. There is a real creation story of humanity, and one only, and it is not a myth. It is being worked out and tested, and enriched and strengthened, step by step.

I will propose that scientific advances, especially those made during the last two decades, are now sufficient for us to address in a coherent manner the questions of where we came from and what we are. To do so, however, we need answers to two even more fundamental questions the query has raised. The first is why advanced social life exists at all, and has occurred so rarely in the history of life. The second is the identity of the driving forces that brought it into existence.

These problems can be solved by bringing together information from multiple disciplines, ranging from molecular genetics, neuroscience, and evolutionary biology to archaeology, ecology, social psychology, and history.

To test any such theory of complex process, it is useful to bring into the light those other social conquerors of Earth, the highly social ants, bees, wasps, and termites, and I will do so. They are needed for perspective in developing the theory of social evolution. I realize I can be easily misinterpreted by putting insects next to people. Apes are bad enough, you might say, but insects? In human biology it is always profitable to make such juxtapositions. There are precedents to comparing the lesser with the greater. Biologists have turned with great success to the bacteria and yeasts to learn the principles of human molecular genetics. They have depended on roundworms and mollusks to learn the basis of our own neural organization and memory. And fruit flies have taught us a great deal about the development of human embryos. We have no less to learn from the social insects, in this case to add background to the origin and meaning of humanity.

II

Where Do We Come From?

· 2 ·

The Two Paths to Conquest

HUMAN BEINGS CREATE cultures by means of malleable languages. We invent symbols that are intended to be understood among ourselves, and we thereby generate networks of communication many orders of magnitude greater than that of any animal. We have conquered the biosphere and laid waste to it like no other species in the history of life. We are unique in what we have wrought.

But we are not unique in our emotions. There are to be found, as in our anatomy and facial expressions, what Darwin called the indelible stamp of our animal ancestry. We are an evolutionary chimera, living on intelligence steered by the demands of animal instinct. This is the reason we are mindlessly dismantling the biosphere and, with it, our own prospects for permanent existence.

Humanity is a magnificent but fragile achievement. Our species is still more impressive because we are the culmination of an evolutionary epic that was continuously played out in great peril. Most of the time our ancestral populations were very small, of a size that in the course of mammalian history typically carried a probability of early extinction. All the prehuman bands taken together made up a population of at most a few tens of thousands of individuals. Very early, the prehuman ancestors split into two or more at a time. During this period the average life of a mammalian species was only half a million years. In conformity to that principle, most of the prehuman collateral lines vanished. The one destined to give rise to modern humanity veered close to extinction itself at least once and possibly many times over the past half million years. The epic might easily have ended at any such constriction, gone forever in a geological eyeblink. It could have happened during a severe drought at the wrong time and place, or an alien disease sweeping into the population from surrounding animals, or pressure from other, more competitive primates. There would then have followed—nothing. The evolution of the biosphere would have pulled back, never again to produce what we became.

The social insects, which currently rule the invertebrate land environment, mostly evolved into existence well over 100 million years ago. Estimates made by specialists are mid-Triassic, or 220 million years ago, for the termites; Late Jurassic to Early Cretaceous, about 150 million years ago for the ants; and for the bumblebees and honeybees, Late Cretaceous, approximately 70–80 million years ago. Thereafter and for the remainder of the Mesozoic era, the diversity of the species in these several evolving lines increased in concert with the rise and spread of the flowering plants. Still, ants and termites acquired their present spectacular dominance among the land-dwelling invertebrates only after they had been around for a long period of time. Their full power was achieved gradually, one innovation at a time, reaching its current levels between 65 and 50 million years ago.

As the swarms of ants and termites spread around the world, many other terrestrial invertebrates coevolved with them and, as a result, not only survived but prospered. Plants and animals evolved defenses against their depredations. Many became specialized to rely on ants, termites, and bees as food. These predators even included pitcher plants, sundews, and other plants able to trap and digest large numbers to add to the nutrients obtained from the soil. A vast array of plant and animal species formed intimate symbioses with the social insects, accepting them as partners. A large percentage came to depend on them entirely for survival, variously as prey, symbionts, scavengers, pollinators, or turners of the soil.

Overall, the pace of evolution of ants and termites was slow enough to be balanced by counterevolution in the rest of life. As a result, these insects were not able to tear down the rest of the terrestrial biosphere by force of numbers, but became vital elements of it. The ecosystems they dominate today are not only sustainable but dependent on them.

In sharp contrast, human beings of the single species *Homo sapiens* emerged in the last several hundred thousand years and spread around the world only during the last sixty thousand years. There was not time for us to coevolve with the rest of the biosphere. Other species were not prepared for the onslaught. This shortfall soon had dire consequences for the rest of life.

At first there was an environmentally benign process of species formation in the populations of our immediate ancestors scattered throughout the Old World. Most led to extinction and hence phylogenetic dead ends—twigs on the tree of life that stopped growing. A zoologist will tell you that there was nothing unusual in this geographical pattern. In the Lesser Sunda archipelago east of Java lived the strange miniature "hobbits," *Homo floresiensis.* They had brains not much larger than those of chimpanzees yet developed stone tools. Of their lives we otherwise know very little. In Europe and the Levant were to be found the Neanderthals, *Homo neanderthalensis,* a sister species of our own *Homo sapiens.* Omnivores like our own ancestors, the Neanderthals had massive bone structures and brains even larger than those of modern *Homo sapiens.* They used crude but nevertheless specialized stone tools. Most of their populations adapted to the harsh climates of the "mammoth steppe," the cold grasslands fringing the continental glacier. They might in time have evolved into an advanced human form of their own, but declined to extinction without further advance. Finally, completing the human beastiary in northern Asia, and known only from a few bone fragments as I write, was another species, the "Denisovans," evidentially vicariant to the Neanderthals occupying land to the east.

None of these species of *Homo*—and let us be generous and call them the other human species—has survived to the present day. Had any done so, it is mind-boggling to think of the moral and religious issues they would have created in modern times. (Civil rights for the Neanderthals? Special education for the hobbits? Salvation and heaven for all?) Although direct evidence is lacking, there can be little doubt about the cause of the extinction of the Neanderthals, which, judging from remains at Gibraltar, was no later than thirty thousand years ago. By one means or another, through competition for food and space or outright slaughter or both, our ancestors were the future exterminators of this and any other species that arose during the adaptive radiation of *Homo*. Isolated in Africa while the Neanderthals still lived were archaic strains of *Homo sapiens*, its descendants destined to expand explosively out of the continent. They populated the Old World all the way to Australia and finally pressed beyond to the New World and distant archipelagoes of Oceania. In the proceess, all other human species encountered were swamped and erased.

Only ten thousand years ago came the invention of agriculture, occurring at least eight times independently in the combined Old and New Worlds. Its adoption dramatically increased the food supply and, with it, the density of people on the land. This decisive advance unleashed exponential population growth and the conversion of most of the natural land environment into drastically simplified ecosystems. Wherever humans saturated wildlands, biodiversity was returned to the paucity of its earliest period half a billion years previously. The rest of the living world could not coevolve fast enough to accommodate the onslaught of a spectacular conqueror that seemed to come from nowhere, and it began to crumble from the pressure.

Even by strictly technical definition as applied to animals, *Homo sapiens* is what biologists call "eusocial," meaning group members containing multiple generations and prone to perform altruistic acts as part of their division of labor. In this respect, they are technically comparable to ants, termites, and other eusocial insects. But let me add immediately: there are major differences between humans and the insects even aside from our unique possession of culture, language, and high intelligence. The most fundamental among them is that all normal members of human societies are capable of reproducing and that most compete with one another to do so.

Also, human groups are formed of highly flexible alliances, not just among family members but between families, genders, classes, and tribes. The bonding is based on cooperation among individuals or groups who know one another and are capable of distributing ownership and status on a personal basis.

The necessity for fine-graded evaluation by alliance members meant that the prehuman ancestors had to achieve eusociality in a radically different way from the instinct-driven insects. The pathway to eusociality was charted by a contest between selection based on the relative success of individuals within groups versus relative success among groups. The strategies of this game were written as a complicated mix of closely calibrated altruism, cooperation, competition, domination, reciprocity, defection, and deceit.

To play the game the human way, it was necessary for the evolving populations to acquire an ever higher degree of intelligence. They had to feel empathy for others, to measure the emotions of friend and enemy alike, to judge the intentions of all of them, and to plan a strategy for personal social interactions. As a result, the human brain became simultaneously highly intelligent and intensely social. It had to build mental scenarios of personal relationships rapidly, both short-term and long-term. Its memories had to travel far into the past to summon old scenarios and far into the future to imagine the consequences of every relationship. Ruling on the alternative plans of action were the amygdala and other emotion-controlling centers of the brain and autonomic nervous system.

Thus was born the human condition, selfish at one time, selfless at another, the two impulses often conflicted. How did *Homo sapiens* reach this unique place in its journey through the great maze of evolution? The answer is that our destiny was foreordained by two biological properties of our distant ancestors: large size and limited mobility.

Back in the Mesozoic era, the first mammals were tiny compared with the largest dinosaurs around them. But they were then, as they remain to this day, mammoth in comparison with insects and other, mostly invertebrate animals. After the passing of the dinosaurs, and as the Age of Reptiles gave way to the Age of Mammals, the mammals proliferated into thousands of species and filled a wide array of niches, from bats in airborne pursuit of flying insects to gigantic plankton-feeding whales plying blue water from pole to pole. The smallest bat is the size of a bumblebee, and the blue whale, growing to eighty feet in length and weighing up to 120 tons, is the largest animal of any kind that has ever lived.

During the adaptive radiation of the mammalian species on the land, a few came to exceed ten kilograms in weight, including deer and other plant-eating animals, along with big cats and other carnivores that preyed on them. It is likely that the number of species worldwide at any given time was between five and ten thousand. Among them appeared the Old World primates, and then, in the Late Eocene period roughly 35 million years ago, the earliest Catarrhini, including species that were to give birth to the present-day Old World monkeys, apes, and humans. Approximately 30 million years ago, the ancestors of the Old World monkeys diverged in evolution from those of modern apes and humans. Some of the proliferating species of the latter specialized on the consumption of plants, others on meat obtained by hunting or scavenging. A few fed on a mix of the two. From one of the branches of mammalian radiation arose the early prehuman line.

For more reasons than size alone, the prehumans were a radically new kind of candidate for eusociality. Insects, from their origin in the first vegetation on land during the Early Devonian 400 million years ago to the present day, have been encased in a knight's armor of chitinous exoskeleton. At the end of each interval of growth, they must create new, more expansive armor and shed the old above it. Whereas the muscles of mammals and other vertebrates are on the outside of the bones, and pull on their outer surface, the muscles of insects are encased by their chitinous skeleton and must pull from the inside. For these reasons insects cannot grow to the size of mammals. The largest among them in the world are African goliath beetles, which are the size of a human fist, and wetas, cricket-like insects of almost equal size that evolved to take the ecological role in New Zealand of mice in the absence of native species on this remote archipelago.

It follows that while eusocial species can dominate the insect world in terms of numbers of individuals, they had to rely on small brains and pure instinct for their conquest. Furthermore, and fundamentally, they were too small to ignite and control fire. They never, no matter how many eons should pass, could have achieved eusociality the human way.

Working their way along the twisting road to eusociality, insects nevertheless had an advantage: they had wings and could travel across the land much farther than mammals. The difference becomes obvious when adjusted to scale. A human band setting out to start a new colony can comfortably travel ten kilometers in a day to emigrate from one campsite to another. A newly inseminated fire ant queen, to take a typical example from among the thousands of species of ants, can fly about the same distance in a few hours to begin a new colony. Upon landing, she breaks off her wings, which are composed of dead tissue (like human hair and fingernails). Then she digs a small nest in the soil, and inside raises a brood of daughter workers from fat and muscle reserves in her own body. A human being is about two hundred times longer than a fire ant queen. So a ten-kilometer flight for an ant is the equivalent of a walk from Boston to Washington, D.C., for a human. Even a half-minute flight of a hundred meters made by a winged ant from her nest of birth to a nest site of her own, is the equivalent of a half-marathon for an earthbound human.

The magnitude of an insect flight results in a far greater scattering of individual queen ants each generation, relative to size. The same would have been true for the solitary wasp ancestors of ants, as well as the solitary protoblattoid ancestors of termites.

The difference between the flying ant ancestors, with each progenitor of the next generation departing on her own, and the plodding mammalian ancestors of humans, which were forced to stay close to others, might seem at first to make the origin of advanced social behavior less likely to evolve in insects. But the opposite is true. In a constantly changing environment, the flying ant is more likely than the wandering mammal to find unoccupied space when she lands. Further, the territory she needs to survive is much smaller than that of a mammal, and is less likely to overlap with already established territories of individuals of the same species.

The potential social insect has another edge: the female colonist needs no male on her journey. Once she has been inseminated during her mating flight, she carries the sperm she receives in a little storage bag (the spermatheca) inside her abdomen. She can pay out one sperm at a time to fertilize her eggs, creating hundreds or thousands of workers over a period of years. Leafcutter ants hold the record: one queen can give birth to 150 million daughter workers during her life span of about a dozen years. Three to five million of these minions are alive at any given time—a size falling between the human populations of Latvia and Norway.

Mammals, especially carnivores, have much larger territories to defend when they settle down to build a nest. Wherever they travel, they are likely to encounter rivals. Females cannot store sperm in their bodies. They must find a male and mate for each parturition. Should the opportunities and pressures of the environment make social grouping profitable, it must be done with personal bonds and alliances based on intelligence and memory.

To summarize to this point on the two social conquerors of Earth, the physiology and life cycle in the ancestors of the social insects and those of humans differed fundamentally in the evolutionary pathways followed to the formation of advanced societies. The insect queen could produce robotic offspring guided by instinct; the prehumans had to rely on bonding and cooperation among individuals. The insects could evolve to eusociality by individual selection in the queen line, generation to generation; the prehumans evolved to eusociality by the interplay of selection at the level of individual selection and at the level of the group.

· 3 ·

The Approach

No individual path of evolution of any kind can be predicted, either at the beginning or even toward the end of its trajectory. Natural selection can bring a species to the brink of a major revolutionary change, only to turn it away. However, some trajectories of evolution can be judged as either possible or impossible, at least on this planet. Insects can evolve to be almost microscopic, but never as big as elephants. Pigs could become aquatic, but their descendants will never fly.

The possible evolution of a species can be visualized as a journey through a maze. As a major advance such as the origin of eusociality is approached, each genetic change, each turn in the maze either makes the attainment of that level less likely, or even impossible, or else keeps it open for access to the next turn. In the earliest steps that keep other options alive, there is still a long way to go, and the ultimate, far distant attainment is least probable. In the last few steps, there is only a short distance to go, and the attainment becomes more probable. The maze itself is subject to evolution along the way. Old corridors (ecological niches) may close, while new ones may open. The structure of the maze depends in part on who is traveling through it, including each of the species.

In every game of evolutionary chance, played from one generation to the next, a very large number of individuals must live and die. The number, however, is not countless. A rough estimate can be made of it, providing at least a plausible order-of-magnitude guess. For the entire course of evolution leading from our primitive mam-

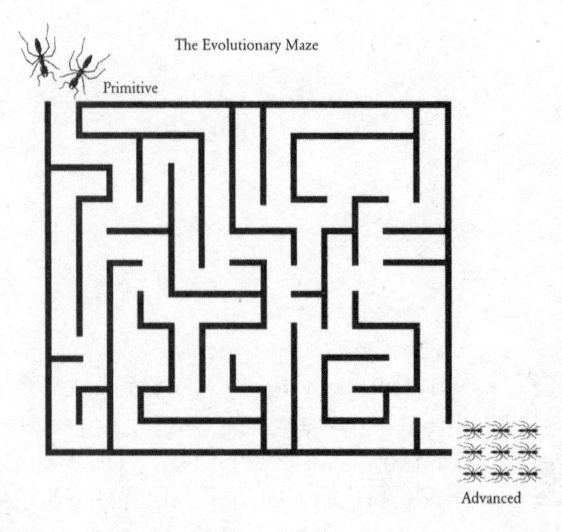

FIGURE 3-1. *The evolution of a species can be visualized as a maze presented by the environment, with opportunities repeatedly closing or remaining open as the maze itself evolves. In the example depicted here, the pathway is from a primitive social life to a highly social one.*

malian forebears of a hundred million years ago to the single lineage that threaded its way to become the first *Homo sapiens*, the total number of individuals it required might have been one hundred billion. Unknowingly, they all lived and died for us.

Many of the players, among the other evolving species, each containing on average a few thousand breeding individuals per generation, also frequently declined and disappeared. Had this happened to any one of the long line of ancestors leading to *Homo sapiens*, the human epic would have promptly ended. Our prehuman ancestors were not chosen, nor were they great. They were just lucky.

Recent research in several disciplines of science is coming together to illuminate the evolutionary steps leading to the human condition, offering at least a partial solution to the "human uniqueness problem" that has so bedeviled science and philosophy. Viewed through time from the beginning to the attainment of the human condition, each step can be interpreted as a preadaptation. In putting it that way, I do not mean to imply that the species leading to our own were in any manner guided to such an end. Rather, each step was an adaptation in its own right—the response of natural selection to conditions prevailing around the species at that place and time.

The first preadaptation was the aforementioned large size and relative immobility that predetermined the trajectory of mammalian evolution, as distinct from that of the social insects. The second preadaptation in the human-bound timeline was the specialization of the early primates, 70 to 80 million years ago, to life in the trees. The most important feature evolved in this change was hands and feet built for grasping. Moreover, their shape and muscles were better suited for swinging from branches, rather than merely grasping them for support. Their efficiency was increased by the simultaneous appearance of opposable thumbs and great toes. It was increased further by modification of the finger and toe tips into flat nails, as opposed to sharp downcurving claws of the kind possessed by most other kinds of arboreal mammals. In addition, the palms and soles were covered by cutaneous ridges that aided in grasping; and they were supplied with pressure receptors that enhanced the sense of touch. Thus equipped, the early primate could use its hand to pick and tease apart pieces of fruit while pulling out individual seeds. The fingernail edges could both cut and scrape objects grasped by the hands. Such an animal, using its hind legs for locomotion, would be able to carry food for considerable distances. It need not use its jaws for that purpose in the manner of a cat or dog. Nor was it required to regurgitate the food to its young like a nesting bird.

Perhaps as an accommodation to the relatively complex manner and flexibility of their feeding behavior, and to the three-dimensional and open vegetation of their habitat, the early prehuman primates evolved a larger brain. For the same reason, they came to depend more on vision and less on smell than did most other mammals. They acquired large eyes with color vision, which were placed forward on the head to give binocular vision and a better sense of

FIGURE 3-2. *A chimpanzee walks bipedally through the savanna forest of Fongoli, Senegal. (From Mary Roach, "Almost Human,"* National Geographic, *April 2008, p. 128. Photograph by Frans Lanting. Frans Lanting / National Geographic Stock.)*

depth. When walking, the prehuman primate did not move its hind legs well apart in parallel; instead, it alternated its legs almost in a single line, one foot placed in front of the other. The offspring, moreover, were fewer in number and required more time to develop.

When one line of these strange arboreal creatures evolved to live on the ground, as it happened in Africa, the next preadaptation was taken—one more fortunate turn in the evolutionary maze. Bipedalism was adopted, freeing the hands for other purposes. The two living species of chimpanzees, the common chimpanzee and bonobo, man's closest phylogenetic relatives, also proceeded far in this direction and at about the same time. On the ground today, they frequently raise their arms and run or walk on their hind legs. They can even make primitive tools.

Following their divergence in evolution from the chimp line, the prehumans, now distinguishable as a group of species called the australopithecines, took the trend to bipedal walking much farther.

FIGURE 3-3. *A chimpanzee sits atop a termite mound in the habitat that gave rise to prehumans. Here they also use crude tools. (From W. C. McGrew, "Savanna chimpanzees dig for food,"* Proceedings of the National Academy of Sciences, U.S.A. *104[49]: 19167–19168 [2007]. Photograph by Paco Bertolani, Leverhulme Centre for Human Evolutionary Studies.)*

Their body as a whole was accordingly refashioned. The legs were lengthened and straightened, and the feet were elongated to create a rocking movement during locomotion. The pelvis was reformed into a shallow bowl to support the viscera, which now pressed toward the legs instead of being slung, ape-like, beneath the horizontal body.

The bipedal revolution was very likely responsible for the overall success of the australopithecine prehumans—at least as measured by the diversity they achieved in body form, jaw musculature, and dentition. During one period, around two million years ago, at least three australopithecine species lived on the African continent. In their body proportions, erect posture, wobbly head perched on top, and elongated hind limbs on which to run and hop, they would have looked at a considerable distance like modern humans. They almost

FIGURE 3-4. Ardipithecus ramidus, *from fossils found at the Middle Awash area of Ethiopia, is at 4.4 million years the oldest known bipedal predecessor of modern humans. It walked on elongated hind legs while retaining long arms suited for a partial life in trees. (From Jamie Shreeve, "The evolutionary road,"* National Geographic, *July 2010, pp. 34–67. Painting by Jon Foster. Jon Foster / National Geographic Stock.)*

certainly traveled in small groups, in the manner of present-day hunter-gatherers. Their brain was no larger than that of a chimpanzee, yet it was from this assemblage that the ancestral species of the earliest *Homo* was eventually to emerge. In evolution, from diversity comes opportunity, the australopiths found.

The ancestral australopiths and their descendant species forming the genus *Homo* lived in an environment conducive to straight-up walking. They never used knuckle walking as practiced by the chimpanzees and other modern apes, with hands curved into fists and employed as forefeet. Walking with arms swinging at the side in the new, australopith manner conferred speed at minimal energy cost, even as it inflicted back and knee problems in addition to the greater risk imposed by balancing the newly heavy globular head on a delicate vertical neck.

For primates whose bodies had been originally crafted for life in the trees, the bipeds could run swiftly. But they could not match the four-legged animals they hunted as prey. Antelopes, zebras, ostriches, and other animals were able to outrun them with ease over short distances. Millions of years of pursuit by lions and other carnivore sprinters had turned prey species into 100-meter champions. If the early humans, however, could not outsprint such animal Olympians, they could at least outlast them in a marathon. At some point, humans became long-distance runners. They needed only to commence a chase and track the prey for mile after mile until it was exhausted and could be overtaken. The prehuman body, thrusting itself off the ball of the foot with each step and holding a steady pace, evolved a high aerobic capacity. In time the body also shed all of its hair, except on the head and pubis and in the pheromone-producing armpits. It added sweat glands everywhere, allowing increased rapid cooling of the naked body surface.

In *Racing the Antelope*, Bernd Heinrich, a distinguished biologist and record-breaking ultradistance runner, has developed the marathon theme at length. He quotes Shawn Found, the 2000 American national champion at twenty-five kilometers, to express the primal

FIGURE 3-5. *Hunting has been a highly adaptive—and dangerous—practice in human prehistory. The inset, part of the Paleolithic paintings of the Lascaux Cave, depicts a gut-shot bison charging a fallen hunter. A raven (a common scavenger that follows hunters) is close by. (The interpretation is by R. Dale Guthrie in* The Nature of Paleolithic Art *[Chicago: University of Chicago Press, 2005].)*

joy of persistence running: "When you experience the run, you . . . relive the hunt. Running is about thirty miles of chasing prey that can outrun you in a sprint, and tracking it down and bringing life back to your village. It's a beautiful thing."

Meanwhile, the forelimbs of the prehuman ancestors were redesigned for flexibility in the manipulation of objects. The arm, especially that of males, became efficient at throwing objects, including stones, and later spears as well, and so for the first time the prehumans could kill at a distance. The advantage this ability gave them during conflict with other, less well-equipped groups must have been enormous.

At least one population of present-day common chimpanzees has developed the ability to throw stones. The behavior appears to be a cultural innovation, perhaps hit upon by a single individual. But it is inconceivable that any chimpanzee could ever match a modern human athlete. None can fling a rock at ninety miles an hour or a spear for nearly the length of a football field. Nor could a juvenile chimp, even if trained, throw an object with the skill of a human child. Early humans had the innate equipment—and likely the tendency also—to use projectiles in capturing prey and repelling enemies. The advantages gained were surely decisive. Spear points and arrowheads are among the earliest artifacts found in archaeological sites.

The environment in which the prehuman epic unfolded was ideal for the production of the first bipeds and their marathoner descendants. During the period of critical evolution, most of sub-Saharan Africa was in a dry epoch, during which the rainforests retreated toward the equatorial belt while shrinking into scattered strongholds in the north. A large part of the continent was covered by savanna woodland alternating with dry forest and grassland. When foraging in open areas, prehumans and *Homo* could stand and peer over low vegetation, to watch for prey and predators intending to make them prey. When threatened, they could run to the shelters of nearby trees. Acacias and other dominant trees were relatively short, and their canopies consisted of branches spread low to the ground and easy to climb—all to the advantage of bipeds. The structure of the environment was similar to that still preserved at Serengeti, Amboseli, Gorongosa, and the other great parks of East Africa. Poets and tourists alike love the feel of this land, far more than they do other habitats of sub-Saharan Africa. They are likely moved, as I will explain later, by an instinct evolved over millions of years by their ancestors in the very same places.

The cradle of humanity was not the deep rainforests with their towering canopies and dark interiors. Nor was it the relatively featureless grasslands and deserts. Rather, humanity was born in the savanna forest, favored by its complex mosaic of different local habitats.

The next step taken on the road to eusociality was the control of fire. Ground fires spreading from lightning strikes are a commonplace in African grasslands and forests today. When they are suppressed, as by the moist soil in forest patches around streams and in easily flooded swales, the woodland undergrowth thickens

FIGURE 3-6. *Bushmen foraging across the grassland of the southern Kalahari. The scene is probably not much different from that commonly occurring in the same area sixty thousand years ago. (From Stephan C. Schuster et al., "Complete Khoisan and Bantu genomes from southern Africa,"* Nature *463: 857, 943–947 [2010]. Photo © Stephan C. Schuster.)*

until it becomes tinder. A lightning strike or the encroachment of a ground fire can then ignite a wildfire, with the flames sweeping through both the ground vegetation and upward to the canopies of surrounding savanna forest. A few animals, especially the young, sick, and old, are trapped and killed. The roving prehumans could not have failed to discover the importance of wildfires as a source of food. Moreover, they found some of the felled animals already cooked, with flesh easy to tear off and eat.

Australian aboriginals have not only harvested such bounty up to the present but also deliberately spread the fires with tree-branch torches. Might prehumans have done the same? There is no way to know how the practice first occurred, but it is certain that early in the history of *Homo* the control of fire became a pivotal event in the zigzag journey to the modern human condition.

The use of fire was on the other hand forever denied to insects and other terrestrial invertebrates. They were physically too small to ignite tinder or carry a flaming object without becoming part of the fuel. It was, of course, also denied aquatic animals regardless of size or prior degree of intelligence of whatever nature. A *Homo sapiens* level of intelligence can arise only on land, whether here on Earth or on any other conceivable planet. Even in the world of fantasy, mermaids and the god Neptune had to evolve on land before they entered their watery domain.

The next step, and the decisive one for the origin of human eusociality, if we accept the evidence from other animals, was the gathering of small groups at campsites. The assemblies were composed of extended families and also, if surviving modern-day hunter-gatherer societies are a guide, included outsider women obtained by exchange for exogamous marriage.

From abundant archaeological evidence, we know that campsites were used by both early African *Homo sapiens* and its sister European species *Homo neanderthalensis*, as well as their common ancestor *Homo erectus*. Hence the practice dates back at least one million years. There is an a priori reason for believing campsites were the crucial adaptation on the path to eusociality: campsites are in essence nests made by human beings. All animal species that have achieved eusociality, without exception, at first built nests that they defended from enemies. They, as did their known antecedents, raised young in the nest, foraged away from it for food, and brought the bounty back to share with others. A variation in the behavior occurs in primitive termites, the ambrosia beetles, and the gall-making aphids and thrips, for which the food is the nest itself. But the basic arrangement, obedient to the biological principle of the primacy of the nest in eusocial evolution, remains the same.

Altricial bird species—those that rear helpless young—have a similar preadaptation. In a few species young adults remain with the parents for a while to help care for their siblings. But no bird species has gone on to evolve full-blown eusocial societies. Possessing only a

FIGURE 3-7. *African wild dogs. (From E. O. Wilson,* Sociobiology *[Cambridge, MA: Harvard University Press, 1975], pp. 510–511. Drawing by Sarah Landry.)*

beak and claws, they have never been equipped to handle tools with any degree of sophistication, or fire at all. Wolves and African wild dogs hunt in coordinated packs in the same manner as chimpanzees and bonobos, and African wild dogs also dig out dens, where one or two females have a large litter. Some pack members hunt and bring a portion of the food to the queen dog and young, while others remain at home as guards. These remarkable canids, although having adopted the rarest and most difficult preadaptation, have not reached full eusociality, with a worker caste or even ape-level intelligence. They cannot make tools. They lack grasping hands and soft-tipped fingers. They remain four-legged, dependent on their carnassial teeth and fur-sheathed claws.

· 4 ·

The Arrival

TWO MILLION YEARS AGO, hominid primates strode upon elongated hind legs across African soil. If we apply the criterion of genetic diversity, measured by hereditary differences in anatomy, they were a success. They had achieved an adaptive radiation, in which multiple species coexisted in time and overlapped at least partially in their respective geographical ranges. Two or three were australopithecines, and at least three were different enough in brain size and dentition to be placed by taxonomists in the newly evolved genus *Homo.* All lived in a complex world of interlaced savanna, savanna forest, and riverine gallery forest. The australopiths were vegetarian, subsisting on a diet of leaves, fruit, underground tubers, and seeds. The *Homo* species also gathered and consumed vegetable food, but in addition they ate meat, most likely by sharing carcasses of larger prey brought down by other predators, as well as by catching smaller animals they could handle themselves. That change, entering one available branch in the evolutionary maze, was to make all the difference.

These hominid primates of two million years ago were diverse, yet no more so than the antelopes and circopithecoid monkeys teeming around them. They were rich in potential—as our own presence bears witness. Nevertheless, from one generation to the next their continued existence was precarious. Their populations were sparse in comparison with the large herbivores, and they were less abundant than some of the human-sized carnivores that hunted them.

During the frequently harsh ten-million-year Neogene period,

FIGURE 4-1. *A reconstruction of a band of* Australopithecus afarensis, *a human predecessor and likely ancestor that lived in Africa five to three million years ago. (© John Sibbick. From* The Complete World of Human Evolution, *by Chris Stringer and Peter Andrews [London: Thames and Hudson, 2005], p. 119.)*

extending before and during the rise of the hominid primates, new mammal species as large as humans evolved more frequently, but they suffered extinction more often as well. Smaller mammals on average were able to buffer themselves better than large mammals, including humans, against extreme environmental changes. Their methods included burrowing, hibernation, and prolonged torpor, adaptations not available to large mammals. Paleontologists have determined that the turnover in species is still higher in mammals that form social groups. They have pointed out that social groups tend to stay apart from each other during breeding, thus creating smaller populations, making them subject to both quicker genetic divergence and higher extinction rates.

During the six-million-year period from the chimpanzee-prehuman divergence to the origin of *Homo sapiens*, fast-moving events occurred that culminated in the breakout of this species from Africa. As continental glaciers advanced south across Eurasia, Africa suffered a period of prolonged drought and cooling. Much of the continent was covered by arid grassland and desert. In these times of stress the death of a few thousand individuals, possibly even just a few hundred, could have snapped the line to *Homo sapiens* altogether. Yet in spite of this environmental gauntlet the hominins were forced to run—or perhaps because of it—*Homo sapiens* emerged, ready to spread out of Africa.

What drove the hominins on through to larger brains, higher intelligence, and thence language-based culture? That, of course, is the question of questions. The australopiths had acquired some of the essential preadaptations. Now one of their species took the further steps that led it to world dominance and the potential of virtually infinite longevity.

That attainment, one of the half dozen great transitions in the history of life, was not made in a simple leap. The evolution that foreshadowed it had begun long before. Between three and two million years ago, one of the australopith species shifted to the consumption of meat. More precisely, it became omnivorous by adding meat to an already existing vegetable diet. The change had occurred by the time of *Homo habilis*, an australopith-derived species known from fossils found at Oldovai Gorge, Tanzania, and dated to 1.8–1.6 million years before the present. Although not definitively identified as the direct ancestor of *Homo sapiens*, *H. habilis* possessed key features that form a link between the primitive australopiths and the earliest known and somewhat more advanced species that can with reasonable certainty be placed as a direct ancestor of *H. sapiens*. The habilines had larger brains than the australopithecines, 640 cubic centimeters in volume as opposed to between 400 and 550 cubic

FIGURE 4-2. *A critical advance in the evolutionary labyrinth.* Homo habilis, *shown here at an imagined kill site, has shifted to a larger dependence on meat, and the use of stone tools to cut up carcasses. (© John Sibbick. From* The Complete World of Human Evolution, *by Chris Stringer and Peter Andrews [London: Thames & Hudson, 2005], p. 133.)*

centimeters, yet still only half that of modern humans (*Homo sapiens*). The molar teeth were reduced in size, a common evolutionary accompaniment of meat consumption. The canines were enlarged, possibly further evidence of the shift to carnivory. The *Homo habilis* skull had thinner brow ridges, and its face projected less forward than that of the more ape-like australopithecines. The folds of the frontal lobe of the brain were arranged in a pattern similar to that of modern humans. Other trends in the brain toward human modernity were well-developed bulges in Broca's area and part of Wernicke's area, a domain of neural centers that organize language in modern humans.

The status of *Homo habilis*, and other hominin species living in Africa between three and two million years ago, is therefore of critical importance in the analysis of human evolution. The changes in the *habilis* skull can be interpreted as the beginning of the evolutionary sprint to the modern human condition. They represent not only an anatomical advance but a basic change in the way of life of the *habilis* population. In simplest terms *habilis* became smarter than the other hominins around them.

Why did one line of australopiths evolve in this direction? A common view held by paleontologists is that changes in the climate and vegetation of Africa favored the evolution of adaptability. Data on the increase and decline of particular animal species indicate that the overall African environment between 2.5 and 1.5 million years ago grew drier. Over most of the continent, rainforests became tropical dry forests and transitional savanna forest, which then turned mostly into continuous grassland and encroaching deserts. The australopithecine ancestors could have adapted to the harsher environment by increasing the variety of their food. They could, for example, have relied on tools to dig up roots and tubers as fallback foods during periods of drought. They surely had the cognitive equipment to do so. In evidence, modern chimpanzees in savanna forest have been observed in this practice, using cow bones and fragments of wood and bark as their digging tools. When near the coast or inland waterways, the australopithecines might also have added shellfish to their diet.

Perhaps, the traditional argument goes, the challenges of new environments gave an advantage to genetic types able to discover and use novel resources to avoid enemies, as well as the capacity to defeat competitors for food and space. Those genetic types were able to innovate and learn from their competitors. They were the survivors of hard times. The flexible species evolved larger brains.

How well does this familiar innovation-adaptiveness hypothesis hold up in studies of other animal species? One analysis made of two hundred bird species introduced by humans into parts of the world outside their native ranges, and hence into alien environments, seems to support the idea. Those species with larger brains relative to their body size were on average better able to establish themselves in the new environments. Further, there is evidence that it was done by greater intelligence and inventiveness. However, the transfer of a documented trend from non-native birds to the human story may be premature. The species studied had been suddenly thrown into radically different environments. The sorting out among them was very different in quality from the natural-selection pressure working on our ancestors among the prehabiline australopiths. Unlike the displaced birds, the prehabilines evolved gradually over many thousands of years with the environment changing around them.

The change that affected the evolution of the early hominids was more likely the increase in the total amount of grassland and savanna forest available to them. The hominids are better conceived of as specialists on those habitats rather than as species adapted to changes occurring around or within the habitats. All naturalists who have worked in savanna forests in particular know the immense variety of sub-habitats that compose these ecosystems. Forest stands of varying density are broken up by swaths of open grassland crossed by riverine woodlands and dotted by copses of dense woods in seasonally flooded swales. Over centuries, individual components change, one giving way to another, back and forth, but the frequency of each and the kaleidoscopic patterns they form together change much more slowly, at least as measured by animal generations and ecological time. As large animals, the hominids must have had home ranges at least ten kilometers in diameter. Among the mixture of habitats present, they could patrol the grassland in search of prey and vegetable food, and race away at the appearance of a predator to nearby copses to climb trees and hide. They could both dig edible tubers in the open ground and collect fruit and edible plant tips from bushes and trees in the woodlands. I suspect they adapted not to one or another of these local sites, or changing from one ecosystem to another, but to the increased area and relative constancy through evolutionary time of the kaleidoscope patterns the sites formed.

It is probable that the early hominids lived in groups of up to several dozen, as do our closest living relatives, the common chimpanzee and bonobo. It may seem self-evident that if complex social behavior requires the evolution of a larger brain proportionate to body size, a larger brain thus suggests the presence of social behavior. If that were true, then a larger brain created by response to a changing environment would be an expected precursor to social behavior. However, when such a relation between brain size and social behavior was tested in a large sample of living and fossil carnivores, including cats, dogs, bears, weasels, and their relatives, no such correlation was found. The association was neither general nor strong enough to create a detectable trend. John A. Finarelli and John J. Flynn, who conducted the research, concluded that "complex processes shaped the modern distribution of encephalization across Carnivora." In other words, multiple selection forces must be sought.

If not adaptation to environmental change (and the matter is far from decided), then what launched the rapid evolutionary growth of the hominid brain? Among the causes, evidenced by the profound changes in the anatomy of the skull and dentition, was likely the shift to a greater reliance on meat as a principal source of protein. This too did not happen suddenly. Prior to the shift, the prehabilines were likely scavenging parts of the carcasses of large animals. The oldest known stone tools, knapped crudely to serve some function or other, date to 6–2 million years before the present. From their oblong shape and sharp edges, and from cut marks found on a fossil antelope bone, it can be reasonably concluded that the tools were used to scavenge meat and marrow from large animals, perhaps after driving other scavengers away to take control. The hominids at this level of evolution were evidently australopiths.

By 1.95 million years ago, during the time of *Homo habilis* and before the appearance of the more modern-looking *Homo erectus*, its descendants, the ancestral hominins, were also taking aquatic prey, including turtles, crocodiles, and fish. The latter were most likely catfish, which even today become densely concentrated in pools during droughts and can be easily caught by hand. In my own zoological field research, I have come upon drought-shrunken ponds where fish and water snakes can be netted and pulled up by the dozens with little effort. (It was so easy that I can imagine myself hunting for dinner with a group of habilines, once they got used to my large size and odd head shape.)

Yet to hunt prey, and thereby obtain animal protein useful for brain development in individual animals, does not of itself explain why the hominid brain grew so dramatically to a huge size. The real cause, it seems, is *how* the prey are hunted. Modern chimpanzees hunt, preying chiefly on monkeys, and obtain about 3 percent of their total calories from meat obtained in this manner. Modern humans, if given a choice, obtain ten times as much. Yet even with their meager incentive, chimpanzees form organized groups and complex strategies when hunting. Their behavior is almost unique among primates. The only other nonhuman primates known to cooperate during hunting are the large-brained capuchin monkeys of Central and South America.

The chimpanzee hunting packs are all-male. They have been observed capturing monkeys in coordinated teams. A monkey that can be separated from its own group is first cornered in a relatively isolated tree. One or two chimpanzees climb the tree to chase the prey down, while others disperse to the bases of adjacent trees to prevent the monkey from traveling to the canopies of other trees and climbing down their trunks to freedom. The prey, when seized, is pummeled and bitten to death. The hunters then tear it apart and share the meat among themselves. Small portions are also passed out, reluctantly, to other members of the troop. The same behavior has been observed in bonobos, the closest living relatives of chimpanzees, but with both sexes participating. The thrill of the hunt is not lost on bonobos, even when dominated by females.

Hunting in groups is rare in mammals as a whole. Other than by primates, it is practiced by lionesses (the one or two males in each pride share in the bounty but seldom hunt themselves). It also occurs in wolves and in African wild dogs.

Chimpanzees and bonobos have an evolutionary history reaching back six million years, the estimated time when their line split from the human clade. We share ancestors before the split, so why have they not also attained the human level? The answer may be the lesser investment the ancestors of chimps and bonobos made in

FIGURE 4-3. Homo erectus, *which research suggests is an immediate ancestor of* Homo sapiens, *took the next two major steps to the modern human social behavior: the establishment of campsites and the control of fires. (© John Sibbick. From* The Complete World of Human Evolution, *by Chris Stringer and Peter Andrews [London: Thames & Hudson, 2005], p. 137.)*

the capture and consumption of live animals. The populations that evolved into *Homo* became specialized for a heavy consumption of animal protein. They needed a high level of teamwork to succeed, and the effort was worth it: meat is gram for gram energetically more efficient than vegetable food. The trend reached an extreme in the populations of *Homo neanderthalensis,* the ice-age sister species of *Homo sapiens,* who depended in winter on hunting animals, including big game.

There remains one piece in the minimal scenario for the emergence of big brains and complex social behavior in the early hominids. Every other kind of animal known that evolved eusociality, as I have stressed, started with a protected nest from which forays can be made to collect food. Other species of relatively large animals that have advanced almost as far as ants into eusociality are the naked mole rats (*Heterocephalus glaber*) of East Africa. They, too, obey the protected-nest principle. Composed of an extended family, each group occupies and defends a system of subterranean burrows. There is a "queen," who is the mother, and "workers," who could

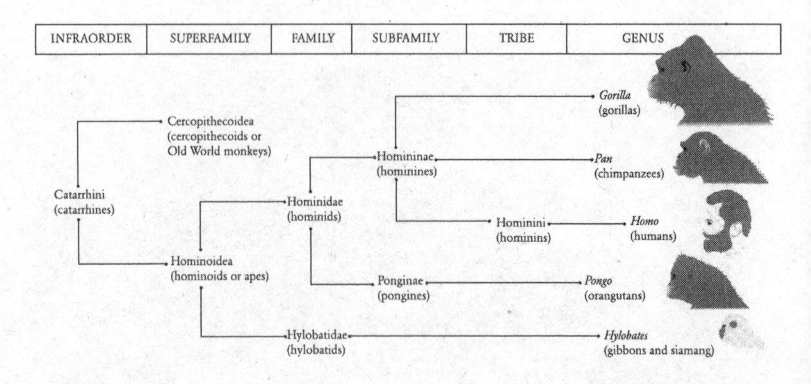

FIGURE 4-4. *The terminology and concept needed to understand human evolution. Depicted here is the branching evolutionary tree of the Old World monkeys and apes, with the scientific and common names of the apes and humans, along with (to the left) names given to each group formed by a major branch. (Modified from Terry Harrison, "Apes among the tangled branches of human origins,"* Science *327: 532–535 [2010]. Reprinted with permission from Harrison [2010].* © Science.*)*

reproduce but do not while the queen remains active. There are even "soldiers," who are most active in defending the nest against snakes and other enemies. A second species, also eusocial but different in details, is the Damaraland mole rat (*Fukomys damarensis*) of Namibia. The closest insect analogs of the naked mole rats are the eusocial thrips and aphids, who stimulate the growth of galls on plants. These hollow swellings are both the insects' nests and source of food.

Why is a protected nest so important? Because members of the group are forced to come together there. Required to explore and forage away from the nest, they must also return. Chimpanzees and bonobos occupy and defend territories, but wander through them while searching for food. The same was probably also true of the australopith and habiline ancestors of man. Chimps and bonobos alternatively break into subgroups and re-aggregate. They advertise the discovery of fruit-laden trees by calling back and forth but do not share the fruit they pick. They occasionally hunt in small packs. Successful members of the pack share the meat among their fellow hunters, but charity mostly comes to an end there. Of greatest importance, the apes have no campfire around which to gather.

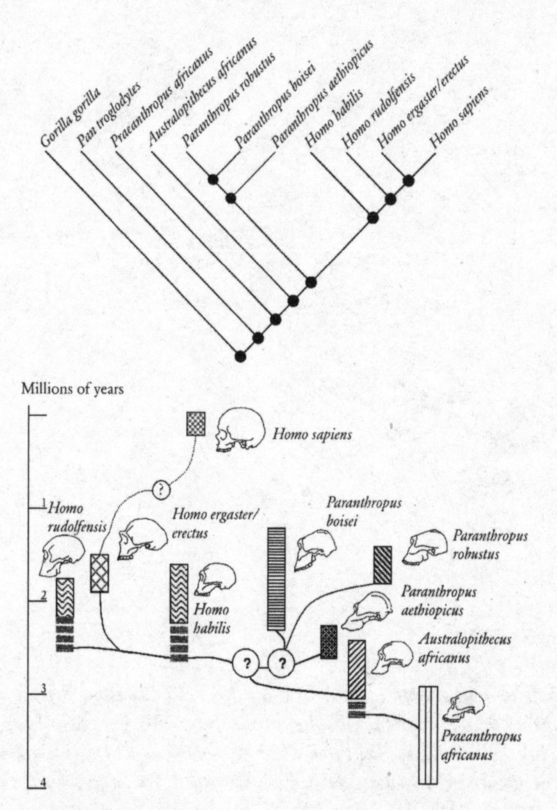

FIGURE 4-5. *The family tree and timeline of the australopiths and primitive* Homo *leading up to the modern human species. (From Winfried Henke, "Human biological evolution," in Franz M. Wuketits and Francisco J. Ayala, eds.,* Handbook of Evolution, *vol. 2,* The Evolution of Living Systems (Including Hominids) *[New York: Wiley-VCH, 2005], p. 167. After D. S. Strait, F. E. Grine, and M. A. Moniz, in* Journal of Human Evolution *32: 17–82 [1997].)*

Carnivores at campsites are forced to behave in ways not needed by wanderers in the field. They must divide labor: some forage and hunt, others guard the campsite and young. They must share food, both vegetable and animal, in ways that are acceptable to all. Otherwise, the bonds that bind them will weaken. Further, the group members inevitably compete with one another, for status of a larger share of food, for access to an available mate, and for a comfortable sleeping place. All of these pressures confer an advantage on those

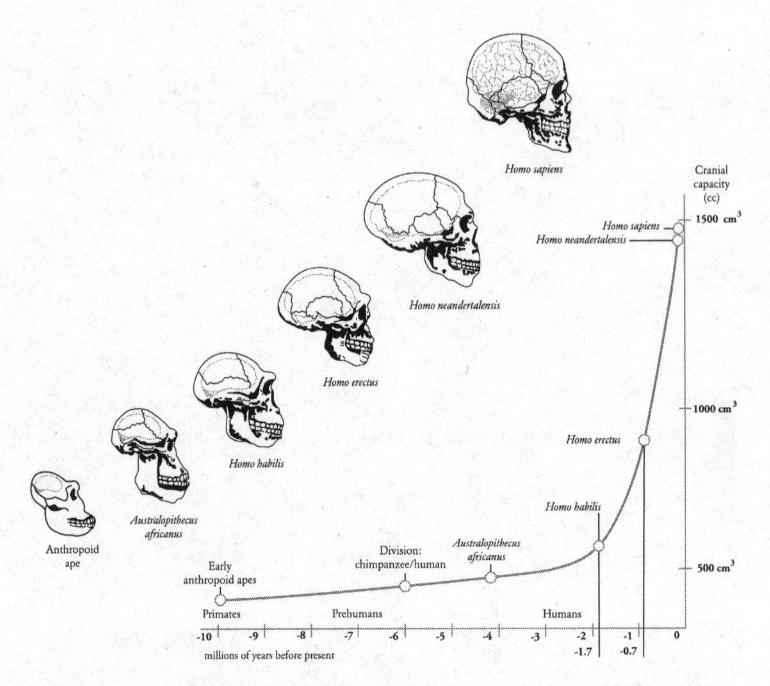

FIGURE 4-6. *The swift growth of the brain that led to its size in modern humanity is depicted here. (Modified from a display in the Exposition Cerveau, Muséum d'Histoire Naturelle de Marseille, France, 22 September to 12 December 2004. © Patrice Prodhomme, Muséum d'Histoire Naturelle d'Aix-en-Provence, France.)*

able to read the intention of others, grow in the ability to gain trust and alliance, and manage rivals. Social intelligence was therefore always at a high premium. A sharp sense of empathy can make a huge difference, and with it an ability to manipulate, to gain cooperation, and to deceive. To put the matter as simply as possible, it pays to be socially smart. Without doubt, a group of smart prehumans could defeat and displace a group of dumb, ignorant prehumans, as true then as it is today for armies, corporations, and football teams.

The cohesion forced by the concentration of groups to protected sites was more than just a step through the evolutionary maze. It was, as I will elaborate later, the event that launched the final drive to modern *Homo sapiens.*

· 5 ·

Threading the Evolutionary Maze

LIKE ALL GREAT PROBLEMS in science, the evolutionary origin of humanity first presented itself as a tangle of partly seen and partly imagined entities and processes. Some of these elements occurred well back in geological time, and may never be understood with certainty. I have nevertheless pieced together those parts of the epic on which I believe researchers agree, and filled in the remainder with informed opinion. The sequence, given in broad strokes, is the consensus I believe to be correct, or at least most consistent with existing evidence.

Overall, it seems now possible to draw a reasonably good explanation of why the human condition is a singularity, why the likes of it has occurred only once and took so long in coming. The reason is simply the extreme improbability of the preadaptations necessary for it to occur at all. Each of these evolutionary steps has been a full-blown adaptation in its own right. Each has required a particular sequence of one or more preadaptations that occurred previously. *Homo sapiens* is the only species of large mammal—thus large enough to evolve a human-sized brain—to have made every one of the required lucky turns in the evolutionary maze.

The first preadaptation was existence on the land. Progress in technology beyond knapped stones and wooden shafts requires fire. No porpoise or octopus, no matter how brilliant, can ever invent a billows and forge. None can ever develop a culture that builds a microscope, deduces the oxidative chemistry of photosynthesis, or photographs the moons of Saturn.

The second preadaptation was a large body size, of a magnitude attained in Earth's history only by a minute percentage of land-dwelling animal species. If an animal at maturity is less than a kilogram in weight, its brain size would be too severely limited for advanced reasoning and culture. Even on land, its body would be unable to make and control fire. That is one reason why leafcutter ants, although the most complex of any species other than humans, and even though they practice agriculture in air-conditioned cities of their own instinctual devising, have made no significant further advance during the twenty million years of their existence.

Next in line of preadaptations was the origin of grasping hands tipped with soft spatulate fingers that were evolved to hold and manipulate detached objects. This is the trait of primates that distinguishes them from all other land-dwelling mammals. Claws and fangs, the ordinary armamentaria of the species, are ill suited for the development of technology. (Writers of Earth-invader science fiction, please remember to provide all your aliens with soft grasping hands or tentacles or some other fleshy fat appendages.)

To use such hands and fingers effectively, candidate species on the path to eusociality had to free them from locomotion in order to manipulate objects easily and skillfully. That was accomplished early by the first prehominids who, as far back as when our presumed ancient forebear *Ardipithecus*, climbed out of the trees, stood up, and began walking entirely on hind legs. Modern humans are geniuses at manipulating things with hands and fingers. We are guided by an extreme development of the kinesthetic sense invested in that ability. The integrative powers of the brain for the sensations that come from handling objects spills out into all other domains of intelligence.

The subsequent step—the next correct turn in the evolutionary maze—was a shift in diet to include a substantial amount of meat, either from scavenged carcasses or from live animals hunted and killed, or both. Meat yields higher energy per gram eaten than does vegetation. Once carnivory is evolutionarily shaped into a niche, less energy is needed to occupy it.

The advantages of cooperation in the harvesting of meat led to the formation of highly organized groups. The earliest societies consisted of extended families but also adoptees and allies. They expanded to a population as large as could be sustained by the local environment. An expanded population was an advantage in the conflicts inevitably arising among different groups. This step and the advantages accruing from it are seen not only in present-day humans—among them both hunter-gatherers and urbanites—but also to a limited extent in chimpanzees.

About a million years ago the controlled use of fire followed, a unique hominid achievement. Firebrands from lightning strikes carried to other sites bestowed enormous advantages on all aspects of our ancestors' existence. Such control improved the yield of meat, allowing more animals to be flushed and trapped. A spreading ground fire was the equivalent of a modern-day pack of hunting dogs. Animals killed in the fire were also often cooked by it. And even in the earliest days of the carnivorous *Homo,* the advantage of meat, sinews, and bone made more easily rendered and consumed had significant consequences. In later evolution, the mastication and physiology of digestion evolved for specialization on cooked meat and vegetables. Cooking became a universal human trait. With the sharing of cooked meals came a universal means of social bonding.

Fire carried about from one place to another was a resource, like meat, fruit, and weapons. Tree limbs and bundles of twigs can smolder for hours. With meat, fire, and cooking, campsites lasting for more than a few days at a time, and thus persistent enough to be guarded as a refuge, marked the next vital step. Such a nest, as it can also be called, has been the precursor to the attainment of eusociality by all other known animals. There is evidence of fossil campsites and their accouterments as far back as *Homo erectus,* the ancestral species intermediate in brain size between *Homo habilis* and modern *Homo sapiens.*

Along with fireside campsites came division of labor. It was spring-loaded: an existing predisposition within groups to self-organize by dominance hierarchies already existed. There were in addition earlier differences between males and females and between young and old. Further, within each subgroup there existed variations in leadership ability, as well as in the proneness to remain at the campsite. The inevitable result emerging quickly out of all these preadaptations was a complex division of labor.

By the time of *Homo erectus*, all of the steps that led this species to eusociality, save the use of controlled fire, had also been followed by modern chimpanzees and bonobos. Thanks to our unique preadaptations, we were ready to leave these distant cousins far behind. The stage was now set for the biggest-brained of African primates to make the truly defining leap to their ultimate potential.

· 6 ·

The Creative Forces

HAD EXTRATERRESTRIAL SCIENTISTS put down on Earth three million years ago, they would have been amazed by the honeybees, mound-building termites, and leafcutter ants, whose colonies were at that time the supreme superorganisms of the insect world and by a wide margin the most complex and ecologically successful social systems on the planet.

The visitors would have also studied the African australopiths, rare bipedal primate species with brains the size of ape brains. Not much potential there or anywhere else among the vertebrate animals, the visitors would surmise. After all, creatures of that size had walked the earth for more than 300 million years past, and nothing much had happened. The eusocial insects seemed the best of which planet Earth was capable.

Imagine further that with their mission accomplished, the extraterrestrials took their leave. Earth's biosphere had stabilized, as far as they could see, and their log would record, "Nothing new of particular importance is likely to happen in the megayears (thousands of millennia) to come. The eusocial insects have been the apex of social evolution for over 100 megayears, and they dominate the terrestrial invertebrate world, and that is likely to continue for another 100 megayears."

However, during their absence, something truly extraordinary happened. The brain of one of the australopiths began to grow rapidly. At the time of the extraterrestrial visit, it measured 500–700 cubic centimeters. By two million years later, it had climbed to 1,000 cubic centimeters. In the next 1.8 million years, it shot on up to 1,500–1,700 cubic centimeters, double that of the ancestral australopithecines. *Homo sapiens* had arrived, and its social conquest of Earth was imminent.

If descendants of the extraterrestrials made a return visit to Earth today, their time in the ensuing three million years having been taken up with more interesting star systems, they would surely be stunned by the situation on Earth. The nearly impossible had happened. One of the bipedal primate species found earlier had not only survived but developed a primitive language-based civilization. And equally surprising, and very disturbing, the primate species was destroying its own biosphere.

Even though tiny in biomass—all of its more than seven billion members could be log-stacked into a cube two kilometers on each edge—the new species had become a geophysical force. They had harnessed the energies of the sun and fossil fuel, diverted a large part of the fresh water for their own use, acidified the ocean, and changed the atmosphere to a potentially lethal state. "It's a terribly botched job of engineering," the visitors might say. "We should have come here earlier and stopped this tragedy from happening."

The origin of modern humanity was a stroke of luck—good for our species for a while, bad for most of the rest of life forever. All of the preadaptations I have cited as evolutionary steps on the road to humanness, if in the right sequence, had the potential to bring a species of big animals to the brink of eusociality. Each of the preadaptations has been cited by one or another scientific author as the key event that catapulted the early hominids to the present human condition. Almost all the conjectures are partially correct. Yet none makes sense except as part of a sequence, one out of many sequences that were possible.

By what *force* of evolutionary dynamics, then, did our lineage thread its way through the evolutionary maze? What in the environment and ancestral circumstance led the species through exactly the right sequence of genetic changes?

The very religious will of course say, the hand of God. That would have been a highly improbable accomplishment even for a supernatural power. In order to bring the human condition into being, a divine Creator would have had to sprinkle an astronomical number of genetic mutations into the genome while engineering the physical and living environments over millions of years to keep the archaic prehumans on track. He might as well have done the same job with a row of random number generators. Natural selection, not design, was the force that threaded this needle.

For almost half a century, it has been popular among serious scientists seeking a naturalistic explanation for the origin of humanity, I among them, to invoke kin selection as a key dynamical force of human evolution. On the surface at least, kin selection, conceived as building a group-level property called inclusive fitness, has been an attractive, even seductive concept. It says parents, offspring, and their cousins and other collateral relatives are bound by the coordination and unity of purpose made possible by selfless acts toward one another. Altruism actually benefits each group member on average because each altruist shares genes by common descent with most other members of its group. Due to the sharing with relatives, its sacrifice increases the relative abundance of these genes in the next generation. If the increase is greater than the average number lost by reducing the number of genes passed on through personal offspring, then the altruism is favored and a society can evolve. Individuals divide themselves into reproductive and nonreproductive castes as a manifestation in part of self-sacrificing behavior on behalf of kin.

Unfortunately for this perception, the foundations of the general theory of inclusive fitness based on the assumptions of kin selection have crumbled, while evidence for it has grown equivocal at best. The beautiful theory never worked well anyway, and now it has collapsed.

A new theory of eusocial evolution, drawn in part from my collaboration with the theoretical biologists Martin Nowak and Corina Tarnita, and in part from the work of other researchers, provides separate accounts for the origin of eusocial insects on the one hand and the origin of human societies on the other. In the case of ants and other eusocial invertebrates, the process is perceived as neither kin selection nor group selection, but individual-level selection, from queen (in the case of ants and other hymenopteran insects) to queen, with the worker caste being an extension of the queen phenotype. Evolution can proceed in this manner because in the early stages of colonial evolution the queen travels far away from her natal colony and creates the members of the colony on her own. The creation of new groups by humans, at the present time and all the way back into prehistory, has been fundamentally different—at least in my personal interpretation and that of some other scientists, when based on comparative biology. Their evolutionary dynamics is driven by both individual and group selection. The multilevel process was first anticipated by Darwin in *The Descent of Man*:

> Now, if some one man in a tribe, more sagacious than the others, invented a new snare or weapon, or other means of attack or defence, the plainest self-interest, without the assistance of much reasoning power, would prompt the other members to imitate him; and all would thus profit. The habitual practice of each new art must likewise in some slight degree strengthen the intellect. If the new invention were an important one, the tribe would increase in number, spread, and supplant other tribes. In a tribe thus rendered more numerous there would always be a rather better chance of the birth of other superior and inventive members. If such men left children to inherit their mental superiority, the chance of the birth of still more ingenious members would be somewhat better, and in a very small tribe decidedly better. Even if they left no children, the tribe would still include their blood-relations; and it has been ascertained by agriculturists that by preserving and breeding from the family of an animal, which when slaughtered was found to be valuable, the desired character has been obtained.

Multilevel selection consists of the interaction between forces of selection that target traits of individual members and other forces of selection that target traits of the group as a whole. The new theory is meant to replace the traditional theory based on pedigree kinship or some comparable measure of genetic relatedness. It has also been provided by Martin Nowak as an alternative to multilevel selection in the case of the social insects. In this approach, it is possible to reduce the entirety of the selective process to its effect on the genome of each colony member and its direct descendants. The result is achieved without reference to the degree of relatedness of each colony, member to members, other than between parent and offspring.

The precursors of *Homo sapiens*, if archaeological evidence and the behavior of modern hunter-gatherers are accepted as guides, formed well-organized groups that competed with one another for territory and other scarce resources. In general, it is to be expected that between-group competition affects the genetic fitness of each member (that is, the proportion of personal offspring it contributes to the group's future membership), whether up or down. A person can die or be disabled, and lose his individual genetic fitness as a result of increased group fitness during, for example, a war or under the rule of an aggressive dictatorship. If we assume that groups are approximately equal to one another in weaponry and other technology, which has been the case for most of the time among primitive societies over hundreds of thousands of years, we can expect that the outcome of between-group competition is determined largely by the details of social behavior within each group in turn. These traits are the size and tightness of the group, and the quality of communication and division of labor among its members. Such traits are heritable to some degree; in other words, variation in them is due in part to differences in genes among the members of the group, hence also among the groups themselves. The genetic fitness of each member, the number of reproducing descendants it leaves, is determined by the cost exacted and benefit gained from its membership in the group. These include the favor or disfavor it earns from other group members on the basis of its behavior. The currency of favor is paid by direct reciprocity and indirect reciprocity, the latter in the form of reputation and trust. How well a group performs depends on how well its members work together, regardless of the degree by which each is individually favored or disfavored within the group.

The genetic fitness of a human being must therefore be a consequence of both individual selection and group selection. But this is true only with reference to the targets of selection. Whether the targets are traits of the individual working in its own interest, or interactive traits among group members in the interest of the group, the ultimate unit affected is the entire genetic code of the individual. If the benefit from group membership falls below that from solitary life, evolution will favor departure or cheating by the individual. Taken far enough, the society will dissolve. If personal benefit from group memberships rises high enough or, alternatively, if selfish leaders can bend the colony to serve their personal interests, the members will be prone to altruism and conformity. Because all normal members have at least the capacity to reproduce, there is an inherent and irremediable conflict in human societies between natural selection at the individual level and natural selection at the group level.

Alleles (the various forms of each gene) that favor survival and reproduction of individual group members at the expense of others are always in conflict with alleles of the same and alleles of other genes favoring altruism and cohesion in determining the survival and reproduction of individuals. Selfishness, cowardice, and unethical competition further the interest of individually selected alleles, while diminishing the proportion of altruistic, group-selected alleles. These destructive propensities are opposed by alleles predisposing individuals toward heroic and altruistic behavior on behalf of members of the same group. Group-selected traits typically take the fiercest degree of resolve during conflicts between rival groups.

It was therefore inevitable that the genetic code prescribing social behavior of modern humans is a chimera. One part prescribes traits that favor success of individuals within the group. The other part prescribes the traits that favor group success in competition with other groups.

Natural selection at the individual level, with strategies evolving that contribute maximum number of mature offspring, has prevailed throughout the history of life. It typically shapes the physiology and behavior of organisms to suit a solitary existence, or at most to membership in loosely organized groups. The origin of eusociality, in which organisms behave in the opposite manner, has been rare in the history of life because group selection must be exceptionally powerful to relax the grip of individual selection. Only then can it modify the conservative effect of individual selection and introduce highly cooperative behavior into the physiology and behavior of the group members.

The ancestors of ants and other hymenopterous eusocial insects (ants, bees, wasps) faced the same problem as those of humans. They finessed it by evolving extreme plasticity of certain genes, programmed so that the altruistic workers have the same genes for physiology and behavior as the mother queen, even though they differ drastically from the queen and among one another in these traits. Selection has remained at the individual level, queen to queen. Yet selection in the insect societies continues at the group level, with colony pitted against colony. This seeming paradox is easily resolved. As far as natural selection in most forms of social behavior is concerned, the colony is operationally only the queen and her phenotypic extension in the form of robot-like assistants. At the same time, group selection promotes genetic diversity among the workers in other parts of the genome to help protect the colony from disease. This diversity is provided by the male with whom each queen mates. In this sense, the genotype of an individual is a genetic chimera. It contains genes that do not vary among colony members, with castes being plastic forms created from the same genes, and genes that do vary among colony members as a shield against disease.

In mammals such a finesse was not possible, because their life cycle is fundamentally different from that of insects. In the key reproductive step of the mammal life cycle, the female is rooted to the territory of her origin. She cannot separate herself from the group in which she was born, unless she crosses over directly to a neighboring group—a common but tightly controlled event in both animals and humans. In contrast, the insect female can be mated, then carry the sperm like a portable male in her spermatheca long distances. She is able to start new colonies all by herself far from the nest of her birth.

The overpowering of individual selection by group selection has not only been rare in mammals and other vertebrates; it has never been and will likely never be complete. The fundamentals of the mammalian life cycle and population structure prevent it. No insect-like social system can be created in the theater of mammalian social evolution.

The expected consequences of this evolutionary process in humans are the following:

- Intense competition occurs between groups, in many circumstances including territorial aggression.
- Group composition is unstable, because of the advantage of increasing group size accruing from immigration, ideological proselytization, and conquest, pitted against the opportunities to gain advantage by usurpation within the group and fission to create new groups.
- An unavoidable and perpetual war exists between honor, virtue, and duty, the products of group selection, on one side, and selfishness, cowardice, and hypocrisy, the products of individual selection, on the other side.
- The perfecting of quick and expert reading of intention in others has been paramount in the evolution of human social behavior.
- Much of culture, including especially the content of the creative arts, has arisen from the inevitable clash of individual selection and group selection.

In summary, the human condition is an endemic turmoil rooted in the evolution processes that created us. The worst in our nature coexists with the best, and so it will ever be. To scrub it out, if such were possible, would make us less than human.

· 7 ·

Tribalism Is a Fundamental Human Trait

TO FORM GROUPS, drawing visceral comfort and pride from familiar fellowship, and to defend the group enthusiastically against rival groups—these are among the absolute universals of human nature and hence of culture.

Once a group has been established with a defined purpose, however, its boundaries are malleable. Families are usually included as subgroups, although they are frequently split by loyalties to other groups. The same is true of allies, recruits, converts, honorary inductees, and traitors from rival groups who have crossed over. Identity and some degree of entitlement are given each member of a group. Conversely, any prestige and wealth he may acquire lends identity and power to his fellow members.

Modern groups are psychologically equivalent to the tribes of ancient history and prehistory. As such, these groups are directly descended from the bands of primitive prehumans. The instinct that binds them together is the biological product of group selection.

People must have a tribe. It gives them a name in addition to their own and social meaning in a chaotic world. It makes the environment less disorienting and dangerous. The social world of each modern human is not a single tribe, but rather a system of interlocking tribes, among which it is often difficult to find a single compass. People savor the company of like-minded friends, and they yearn to be in one of the best—a combat marine regiment, perhaps, an elite college, the executive committee of a company, a religious sect, a fraternity, a garden club—any collectivity that can be compared favorably with other, competing groups of the same category.

People around the world today, growing cautious of war and fearful of its consequences, have turned increasingly to its moral equivalent in team sports. Their thirst for group membership and superiority of their group can be satisfied with victory by their warriors in clashes on ritualized battlefields. Like the cheerful and well-dressed citizens of Washington, D.C., who came out to witness the First Battle of Bull Run during the Civil War, they anticipate the experience with relish. The fans are lifted by seeing the uniforms and symbols and battle gear of the team, the championship cups and banners on display, the dancing seminude maidens appropriately called cheerleaders. Some of the fans wear bizarre costumes and face makeup in homage to their team. They attend triumphant galas after victories. Many, especially of warrior and maiden age, shed all restraint to join in the spirit of the battle and the joyous mayhem afterward. When the Boston Celtics defeated the Los Angeles Lakers for the National Basketball Association championship, on a June night in 1984, the team was ecstatic, and the mantra was "Celts Supreme!" The social psychologist Roger Brown, who witnessed the aftermath, commented, "It was not just the players who felt supreme but all their fans. There was ecstasy in the North End. The fans burst out of the Garden and nearby bars, practically break dancing in the air, stogies lit, arms uplifted, voices screaming. The hood of a car was flattened, about thirty people jubilantly piled aboard, and the driver—a fan—smiled happily. An improvised slow parade of honking cars circled through the neighborhood. It did not seem to me that those fans were just sympathizing or empathizing with their team. They personally were flying high. On that night each fan's self-esteem felt supreme; a social identity did a lot for many personal identities."

Brown then added an important point: "Identification with a sports team has in it something of the arbitrariness of the minimal groups. To be a Celtic fan you need not be born in Boston or even live there, and the same is true of membership on the team. As individuals, or with other group memberships salient, both fans and team members might be very hostile. So long as the Celtic membership was salient, however, all rode the waves together."

Experiments conducted over many years by social psychologists have revealed how swiftly and decisively people divide into groups, and then discriminate in favor of the one to which they belong. Even when the experimenters created the groups arbitrarily, then labeled them so the members could identify themselves, and even when the interactions prescribed were trivial, prejudice quickly established itself. Whether groups played for pennies or identified themselves groupishly as preferring some abstract painter to another, the participants always ranked the out-group below the in-group. They judged their "opponents" to be less likable, less fair, less trustworthy, less competent. The prejudices asserted themselves even when the subjects were told the in-groups and out-groups had been chosen arbitrarily. In one such series of trials, subjects were asked to divide piles of chips among anonymous members of the two groups, and the same response followed. Strong favoritism was consistently shown to those labeled simply as an in-group, even with no other incentive and no previous contact.

In its power and universality, the tendency to form groups and then favor in-group members has the earmarks of instinct. It could be argued that in-group bias is conditioned by early training to affiliate with family members and by encouragement to play with neighboring children. But even if such experience does play a role, it would be an example of what psychologists call prepared learning, the inborn propensity to learn something swiftly and decisively. If the propensity toward in-group bias has all these criteria, it is likely to be inherited and, if so, can be reasonably supposed to have arisen through evolution by natural selection. Other cogent examples of prepared learning in the human repertoire include language, incest avoidance, and the acquisition of phobias.

If groupist behavior is truly an instinct expressed by inherited prepared learning, we might expect to find signs of it even in very young children. And exactly this phenomenon has been discovered by cognitive psychologists. Newborn infants are most sensitive to the first sounds they hear, to their mother's face, and to the sounds of their native language. Later they look preferentially at persons who previously spoke their native language within their hearing. Preschool children tend to select native-language speakers as friends. The preferences begin before the comprehension of the meaning of speech and are displayed even when speech with different accents is fully comprehended.

The elementary drive to form and take deep pleasure from in-group membership easily translates at a higher level into tribalism. People are prone to ethnocentrism. It is an uncomfortable fact that even when given a guilt-free choice, individuals prefer the company of others of the same race, nation, clan, and religion. They trust them more, relax with them better in business and social events, and prefer them more often than not as marriage partners. They are quicker to anger at evidence that an out-group is behaving unfairly or receiving undeserved rewards. And they grow hostile to any out-group encroaching upon the territory or resources of their in-group. Literature and history are strewn with accounts of what happens at the extreme, as in the following from Judges 12: 5–6 in the Old Testament:

> The Gileadites captured the fords of the Jordan leading to Ephraim, and whenever a survivor of Ephraim said, "Let me go over," the men of Gilead asked him, "Are you an Ephraimite?" If he replied, "No," they said, "All right, say 'Shibboleth.' " If he said, "Sibboleth," because he could not pronounce the word correctly, they seized him and killed him at the fords of the Jordan. Forty-two thousand Ephraimites were killed at that time.

When in experiments black and white Americans were flashed pictures of the other race, their amygdalas, the brain's center of fear and anger, were activated so quickly and subtly that the conscious centers of the brain were unaware of the response. The subject, in effect, could not help himself. When, on the other hand, appropriate contexts were added—say, the approaching black was a doctor and the white his patient—two other sites of the brain integrated with the higher learning centers, the cingulate cortex and the dorsolateral preferential cortex, lit up, silencing input through the amygdala.

Thus different parts of the brain have evolved by group selection to create groupishness. They mediate the hardwired propensity to downgrade other-group members, or else in opposition to subdue its immediate, autonomic effects. There is little or no guilt in the pleasure experienced from watching violent sporting events and war films, providing the amygdala rules the action and the story unwinds to a satisfying destruction of the enemy.

· 8 ·

War as Humanity's Hereditary Curse

"HISTORY IS A bath of blood," wrote William James, whose 1906 antiwar essay is arguably the best ever written on the subject. "Modern war is so expensive," he continued, "that we feel trade to be a better avenue to plunder; but modern man inherits all the innate pugnacity and all the love of glory of his ancestors. Showing war's irrationality and horror is of no effect on him. The horrors make the fascination. War is the *strong* life; it is life *in extremis*; war taxes are the only ones men never hesitate to pay, as the budgets of all nations show us."

Our bloody nature, it can now be argued in the context of modern biology, is ingrained because group-versus-group was a principal driving force that made us what we are. In prehistory, group selection lifted the hominids that became territorial carnivores to heights of solidarity, to genius, to enterprise. And to *fear.* Each tribe knew with justification that if it was not armed and ready, its very existence was imperiled. Throughout history, the escalation of a large part of technology has had combat as its central purpose. Today, the calendars of nations are punctuated by holidays to celebrate wars won and to perform memorial services for those who died waging them. Public support is best fired up by appeal to the emotions of deadly combat, over which the amygdala is grandmaster. We find ourselves in the *battle* to stem an oil spill, the *fight* to tame inflation, the *war* against cancer. Wherever there is an enemy, animate or inanimate, there must be a victory. You must prevail at the front, no matter how high the cost at home.

Any excuse for a real war will do, so long as it is seen as necessary to protect the tribe. The remembrance of past horrors has no effect. From April to June in 1994, killers from the Hutu majority in Rwanda set out to exterminate the Tutsi minority, which at that time ruled the country. In a hundred days of unrestrained slaughter by knife and gun, 800,000 people died, mostly Tutsi. The total Rwandan population was reduced by 10 percent. When a halt was finally called, two million Hutu fled the country, fearing retribution. The immediate causes for the bloodbath were political and social grievances, but they all stemmed from one root cause: Rwanda was the most overcrowded country in Africa. For a relentlessly growing population, the per capita arable land was shrinking toward its limit. The deadly argument was over which tribe would own and control the whole of it.

The Tutsi had been dominant before the genocide. The Belgian colonists had considered them the better of the two tribes and favored them accordingly. The Tutsi, of course, held the same belief, and although the tribes spoke the same language, they treated the Hutu as inferiors. For their part, the Hutu thought of the Tutsi as invaders who had come generations earlier from Ethiopia. Many of those who attacked their neighbors were promised the land of the Tutsi they killed. When they threw Tutsi bodies into the river, they jeered that they were returning their victims to Ethiopia.

Once a group has been split off and sufficiently dehumanized, any brutality can be justified, at any level, and at any size of the victimized group up to and including race and nation. Russia's Great Terror under Stalin resulted in the deliberate starvation to death of more than three million Soviet Ukrainians during the winter of 1932–33. In 1937 and 1938, 681,692 executions were carried out for alleged "political crimes," of which more than 90 percent were peasants considered resistant to collectivization. The U.S.S.R. as a whole soon itself suffered equally from the brutal Nazi invasion, the stated purpose of which was to subdue the "inferior" Slavs and make room for expansion of the racially "pure" Aryan peoples.

If no other reason is convenient for waging a war of territorial expansion, there has always been God. It was the will of God that brought the Crusaders to the Levant. They were paid in advance with papal indulgences. They marched under the sign of the cross, and demanded that the alleged true Cross be returned to Christian hands. During the siege of Acre in 1191, Richard I brought 2,700 Muslim prisoners of war close enough to the battle line for Saladin to see them, then slaughtered the lot by sword. His motive is said to have been to impress the Moslem leader of the English monarch's iron will, but it could equally have been Richard's wish to keep the prisoners from returning to arms. No matter: the ultimate motivation for all the horror was to wrest land and resources from the Muslims and pass them over to the kingdoms of Christendom.

Then came Islam's turn. It was equally in the service of God that the siege of Constantinople was conducted by the Ottoman Turks under Sultan Mehmed II in 1453. It was the Holy Trinity and all the saints to whom Christians prayed as they huddled in the great church of Hagia Sofia while the Ottoman forces converged upon the Augusteum. The desperate supplicants were not heard. The Moslems were favored by God that day, and so the Christians were variously butchered and sold into slavery.

No one has expressed the deep linkage within the Abrahamic religions between human and divine violence more vividly than Martin Luther in his 1526 essay *Whether Soldiers, Too, Can Be Saved.*

> But what are you going to do about the fact that people will not keep the peace, but rob, steal, kill, outrage women and children, and take away property and honor? The small lack of peace called war or the sword must set a limit to this universal, worldwide lack of peace which would destroy everyone. This is why God honors the sword so highly that he says that he himself has instituted it (Rom. 13:1) and does not want men to say or think that they have invented it or instituted it. For the hand that wields this sword and kills with it is not man's hand, but God's; and it is not man, but God, who hangs, tortures, beheads, kills, and fights. All these are God's works and judgments.

And so it has ever been. According to Thucydides, the Athenians asked the independent people of Melos to abandon their support of Sparta in the Peloponnesian War and submit to Athenian rule. Envoys from the two states met to discuss the issue. The Athenians explained the fate given by the gods to men: "The powerful exact what they can, and the weak grant what they must." The Melians responded that they would never be made slaves and would appeal to the gods for divine justice. The Athenians replied, "Of the gods we believe and of men we know, by a law of their nature, whenever they can rule they will. This law was not made by us, and we are not the first to have acted upon it. We did but inherit it, and we know that you and all of mankind, if you were as strong as we are, would do as we do. So much for the gods. We have told you why we expect to stand as high in their good opinion as you." The Melians still refused, and an Athenian force soon arrived to conquer Melos. In the calm tone of classic Greek tragedy, Thucydides reports, "The Athenians thereupon put to death all who were of military age, and made slaves of the women and children. They then colonized the island, sending thither five hundred settlers of their own."

A familiar fable is told to symbolize this pitiless dark angel of human nature. A scorpion asks a frog to ferry it across a stream. The frog at first refuses, saying that it fears the scorpion will sting it. The scorpion assures the frog it will do no such thing. After all, it says, we will both perish if I sting you. The frog consents, and halfway across the stream the scorpion stings it. Why did you do that, the frog asks as they both sink beneath the surface. It is my nature, the scorpion explains.

It should not be thought that war, often accompanied by genocide, is a cultural artifact of a few societies. Nor has it been an aberration of history, a result of the growing pains of our species' maturation. Wars and genocide have been universal and eternal, respecting no particular time or culture. Since the end of the Second World War, violent conflict between states has declined dras-

FIGURE 8-1. *For the Mayans, war was a regular way of life, as illustrated in the murals at Bonampak, Mexico, about AD 800. (From Thomas Hayden, "The roots of war,"* U.S. News & World Report, *26 April 2004, pp. 44–50. Photograph by Enrico Ferorelli, computer reconstruction by Doug Stern. National Geographic Stock.)*

tically, owing in part to the nuclear standoff of the major powers (two scorpions in a bottle writ large). But civil wars, insurgencies, and state-sponsored terrorism continue unabated. Overall, big wars have been replaced around the world by small wars of the kind and magnitude more typical of hunter-gatherer and primitively agricultural societies. Civilized societies have tried to eliminate torture, execution, and the murder of civilians, but those fighting little wars do not comply.

Archaeological sites are strewn with the evidence of mass conflict. A large part of the most impressive constructions of history have had a defensive purpose, including the Great Wall of China, Hadri-

FIGURE 8-2. *The Yanomamo are one of the last primitive tribes of South America, with a population of ten thousand divided among 200–250 fiercely independent villages. Raids on neighboring villages are commonplace. Here, warriors line up at dawn prior to departing on such a raid, their faces and bodies decorated with masticated charcoal. (Provided with permission to reproduce by Napoleon A. Chagnon.)*

an's Wall across England, the magnificent castles and fortresses of Europe and Japan, the cliff dwellings of the Ancestral Pueblo, the city walls of Jerusalem and Constantinople. Even the Acropolis was originally a walled fortress town.

Archaeologists have found burials of massacred people to be a commonplace. Tools from the earliest Neolithic period include instruments clearly designed for fighting. The Iceman, a frozen body discovered in the Alps in 1991 and determined to be over five thousand years old, died of an arrowhead found embedded in his chest. He carried a bow, a quiver of arrows, and a copper dagger or knife, conceivably for the hunting and dressing of game. But he also possessed a hatchet with a copper blade unmarked by evidence of use by a woodsman with a need to chop wood and bone. More likely it was intended to be a battle-ax.

It is often said that a few surviving hunter-gatherer societies, most notably the Bushmen of South Africa and the Australian Aboriginals, which are close in social organization to our hunter-gatherer ancestors, conduct no wars and therefore bear witness to the late appearance in history of violent mass conflict. But their existence has been marginalized and reduced by European colonists and, in the case of the Bushmen, also by earlier Zulu and Herero invaders. Once the Bushmen lived in larger populations over much wider and more productive habitats than the scrubland and desert they occupy today. They also engaged in tribal wars. Evidence from rock drawings and the accounts of early European explorers and settlers depict pitched battles between armed groups. When the Herero began to invade Bushman territory in the 1800s, they were at first driven out by Bushman war parties.

One might think that the influence of pacific Eastern religions, especially Buddhism, has been consistent in opposing violence. Such is not the case. Whenever Buddhism dominated and became the official ideology, whether Theravāda Buddhism in Southeast Asia or Tantric Buddhism in East Asia and Tibet, war was tolerated and even pressed as part of faith-based state policy. The rationale is simple, and has its mirror image in Christianity: peace, nonviolence, and brotherly love are core values, but a threat to Buddhist law and civilization is an evil that must be defeated. In effect, "Kill them all, and Buddha will receive his own."

In the sixth century Chinese rebels, under the Buddhist title "Greater Vehicle" (Mahāyāna), set out to eliminate all the world's "demons"—starting with the Buddhist clergy. In Japan, Buddhism was modified as an instrument of feudal struggles, creating the hybrid "warrior monk." Only at the end of the sixteenth century were the powerful monasteries broken by the central military gov-

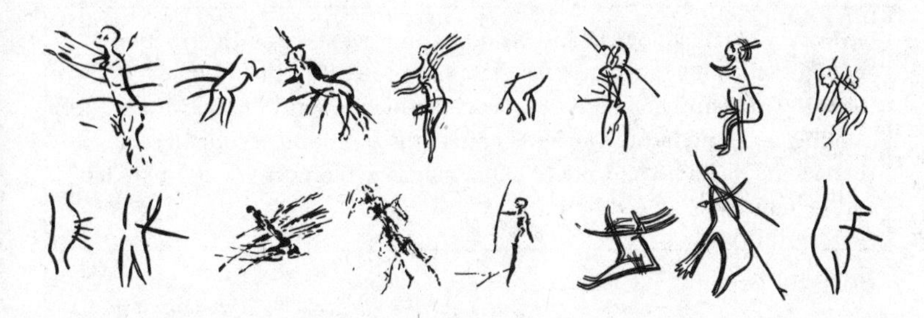

FIGURE 8-3. *Killings of humans by spear thrusts, mostly multiple in nature, are found in the Paleolithic art of various European caves. The mortal wounds could be the result of murder or executions, but they are more likely (in the present author's opinion) to represent enemies felled by war parties that attacked individuals. (From R. Dale Guthrie,* The Nature of Paleolithic Art *[Chicago: University of Chicago Press, 2005].)*

ernment. Buddhism was then modified as an instrument of feudal struggles. After the Meiji Restoration in 1818, Japanese Buddhism became part of the nation's "spiritual mobilization."

And what of distant prehistory? Might warfare be in some manner a consequence of the spread of agriculture and villages and a rising density in people? Such was evidently not the case. Burial sites of foraging people of the Upper Paleolithic and Mesolithic of the Nile Valley and Bavaria include mass interments of what appear to be entire clans. Many had died violently by bludgeon, spear, or arrow. From the Upper Paleolithic 40,000 to about 12,000 years ago, scattered remains often bear evidence of death by blows to the head and cut marks on bones. This was the period of the famous Lascaux and other cave paintings, some of which include drawings of people being speared or lying about already dead or dying.

There is another way to test the prevalence of violent group conflict in deep human history. Archaeologists have determined that after populations of *Homo sapiens* began to spread out of Africa approximately 60,000 years ago, the first wave reached as far as New Guinea and Australia. The descendants of the pioneers remained in these outliers as hunter-gatherers or at most primitive agriculturalists, until

TABLE 8-1. Archaeological and ethnographic evidence on the fraction of adult mortality due to warfare. "Before present" in the middle heading indicates before 2008. [From Samuel Bowles, "Did warfare among ancestral hunter-gatherers affect the evolution of human social behaviors," *Science* 324: 1295 (2009). Primary references are not included in the table reproduced here.]

Site	*Archaeological evidence approx. date (years before present)*	*Fraction of adult mortality due to warfare*
British Columbia (30 sites)	5500–334	0.23
Nubia (site 117)	14–12000	0.46
Nubia (near site 117)	14–12000	0.03
Vasiliv´ka III, Ukraine	11000	0.21
Volos´ke, Ukraine	"Epipalaeolithic"	0.22
S. California (28 sites)	5500–628	0.06
Central California	3500–500	0.05
Sweden (Skateholm 1)	6100	0.07
Central California	2415–1773	0.08
Sarai Nahar Rai, N. India	3140–2854	0.30
Central California (2 sites)	2240–238	0.04
Gobero, Niger	16,000–8200	0.00
Calumnata, Algeria	8300–7300	0.04
Ile Teviec, France	6600	0.12
Bogebakken, Denmark	6300–5800	0.12

Population, region	*Ethnographic evidence (dates)*	*Fraction of adult mortality due to warfare*
Ache, Eastern Paraguay*	Precontact (1970)	0.30
Hiwi, Venezuela-Colombia*	Precontact (1960)	0.17
Murngin, NE Australia*†	1910–1930	0.21
Ayoreo, Bolivia-Paraguay‡	1920–1979	0.15
Tiwi, N. Australia§	1893–1903	0.10
Modoc, N. California§	"Aboriginal times"	0.13
Casiguran Agta, Philippines*	1936–1950	0.05
Anbara, N. Australia*†‖	1950–1960	0.04

*Foragers. †Maritime. ‡Seasonal forager-horticulturalists.
§Sedentary hunter-gatherers. ‖Recently settled.

reached by Europeans. Living populations of similar early provenance and archaic cultures are the aboriginals of Little Andaman Island off the east coast of India, the Mbuti Pygmies of Central Africa, and the !Kung Bushmen of southern Africa. All today, or at least within historical memory, have exhibited aggressive territorial behavior.

Among the very small percentage of the thousand cultures worldwide studied by anthropologists and considered "peaceful" are the Copper and Ingalik Eskimo, the Gebusi of lowland New Guinea, the Semang of peninsular Malaysia, the Amazonian Sirionó, the Yahgan of Tierra del Fuego, the Warrau of eastern Venezuela, and the aborigines of the Tasmanian western coast. At least some had high homicide rates. In the New Guinea Gebusi and Copper Eskimo, a third of all adult deaths were homicides. "This might be explained," the anthropologists Steven A. LeBlanc and Katherine E. Register have written, "by the fact that among small societies almost everyone is a relative, albeit a distant one. Naturally, this raises some perplexing questions: Who is a member of the group and who is an outsider? Which killing is considered a homicide and which killing is an act of warfare? Such questions and answers become somewhat fuzzy. So some of this so-called peacefulness is more dependent on the definition of homicide and warfare than on reality. In fact, some of these societies did have warfare, but it has usually been considered to be minor and insignificant."

The key question remaining in the dynamics of human genetic evolution is whether natural selection at the group level has been strong enough to overcome the powerful force of natural selection at the level of the individual. Put another way, have the forces favoring instinctive altruistic behavior to other members of the group been strong enough to disfavor individual selfish behavior? Mathematical models constructed in the 1970s showed that group selection can prevail if the relative rate of group extinction or diminishment in groups without altruistic genes is very high. As one class of such models suggests, when the rate of increase of group multiplication with altruistic members exceeds the rate of increase of selfish individuals within the groups, gene-based altruism can spread through the population of groups. More recently, in 2009, the theoretical biologist Samuel Bowles has produced a more realistic model that fits the empirical data well. His approach answers the following question: if cooperative groups were more likely to prevail in conflicts with other groups, has the level of intergroup violence been sufficient to influence the evolution of human social behavior? The estimates of adult mortality in hunter-gatherer groups from the beginning of Neolithic times to the present, shown in the accompanying table, support that proposition.

Tribal aggressiveness thus goes well back beyond Neolithic times, but no one as yet can say exactly how far. It could have begun at the time of *Homo habilis*, with a heavy dependence of the populations on scavenging or hunting for meat. And there is a good chance that it could be a much older heritage, dating beyond the split six million years ago between the lines leading to modern chimpanzees and to humans, respectively. A series of researchers, starting with Jane Goodall, have documented the murders within chimpanzee groups and lethal raids conducted between groups. It turns out that chimpanzees and human hunter-gatherers and primitive farmers have about the same rates of death due to violent attacks within and between groups. But nonlethal violence is far higher in the chimps, occurring between a hundred and possibly a thousand times more often than in humans.

Chimpanzees live in groups, called by primatologists "communities," of up to 150 individuals, which defend territories of up to 38 square kilometers, and at low population densities of about 5 individuals per square kilometer. Within each of these assemblages, small parties form into subgroups. The members of each subgroup, averaging 5 to 10 strong, travel, feed, and sleep together. Males spend their entire lives with the same community, whereas most females emigrate when young to join neighboring communities. Males are more gregarious than females. They are also intensely status conscious, frequently engaging in displays that lead to fighting. They form coalitions with others and use a wide array of maneuvers and deceptions to exploit or altogether evade the dominance order. The patterns of collective violence in which young chimp males engage are remarkably similar to those of young human males. Aside from constantly vying for status, both for themselves and for their gangs, they tend to avoid open mass confrontations with rival troops, instead relying on surprise attacks.

The purpose of raids made by the male gangs on neighboring communities is evidently to kill or drive out its members and acquire new territory. The entirety of such conquest under fully natural conditions has been witnessed by John Mitani and his collaborators in Uganda's Kibale National Park. The war, conducted over ten years, was eerily human-like. Every ten to fourteen days, patrols of up to twenty males penetrated enemy territory, moving quietly in single file, scanning the terrain from ground to the treetops, and halting cautiously at every surrounding noise. If a force larger than their own was encountered, the invaders broke rank and ran back to their own territory. When they encountered a lone male, however, they piled on him in a crowd and pummeled and bit him to death. When a female was encountered, they usually let her go. This latter tolerance was not a display of gallantry. If she carried an infant, they took it from her and killed and ate it. Finally, after such constant pressure for so long, the invading gangs simply annexed the enemy territory, adding 22 percent to the land owned by their own community.

There is no certain way to decide on the basis of existing knowledge whether chimpanzee and humans inherited their pattern of territorial aggression from a common ancestor or whether they evolved it independently in response to parallel pressures of natural selection and opportunities encountered in the African homeland. From the remarkable similarity in behavioral detail between the two species, however, and if we use the fewest assumptions required to explain it, a common ancestry seems the more likely choice.

The principles of population ecology allow us to explore more deeply the roots of the origin of mankind's tribal instinct. Population growth is exponential. When each individual in a population is replaced in every succeeding generation by more than one—even by a very slight fraction more, say 1.01—the population grows faster and faster, in the manner of a savings account or debt. A population of chimpanzees or humans is always prone to grow exponentially when resources are abundant, but after a few generations even in the best of times it is forced to slow down. Something begins to intervene, and in time the population reaches its peak, then remains steady, or else oscillates up and down. Occasionally it crashes, and the species becomes locally extinct.

What is the "something"? It can be anything in nature that moves up or down in effectiveness with the size of the population. Wolves, for example, are the limiting factor for the population of elk and moose they kill and eat. As the wolves multiply, the populations of elk and moose stop growing or decline. In parallel manner, the quantity of elk and moose are the limiting factor for the wolves:

when the predator population runs low on food, in this case elk and moose, its population falls. In other instances, the same relation holds for disease organisms and the hosts they infect. As the host population increases, and the populations grow larger and denser, the parasite population increases with it. In history diseases have often swept through the land, called an epidemic in humans and an epizootic in animals, until the host populations decline enough or a sufficient percentage of its members acquire immunity. Disease organisms can be defined as predators that eat their prey in units of less than one.

There is another principle at work: limiting factors work in hierarchies. Suppose that the primary limiting factor is removed for elk by humans killing the wolves. As a result the elk and moose grow more numerous—until the next factor kicks in. The factor may be that herbivores overgraze their range and run short of food. Another limiting factor is emigration, where individuals have a better chance to survive if they leave and go someplace else. Emigration due to population pressure is a highly developed instinct in lemmings, plague locusts, monarch butterflies, and wolves. If such populations are prevented from emigrating, the populations might again increase in size, but then some other limiting factor manifests itself. For many kinds of animals, the factor is the defense of territory, which protects the food supply for the territory owner. Lions roar, wolves howl, and birds sing in order to announce that they are in their territories and desire competing members of the same species to stay away. Humans and chimpanzees are intensely territorial. That is the apparent population control hardwired into their social systems. What the events were that occurred in the origin of the chimpanzee and human lines—before the chimpanzee-human split of six million years ago—can only be speculated. I believe, however, that the evidence best fits the following sequence. The original limiting factor, which intensified with the introduction of group hunting for animal protein, was food. Territorial behavior evolved as a device to sequester the food supply. Expansive wars and annexation resulted in enlarged territories and favored genes that prescribe group cohesion, networking, and the formation of alliances.

For hundreds of millennia, the territorial imperative gave stability to the small, scattered communities of *Homo sapiens*, just as they do today in the small, scattered populations of surviving hunter-gatherers. During this long period, randomly spaced extremes in the environment alternately increased and decreased the population size that could be contained within territories. These "demographic shocks" led to forced emigration or aggressive expansion of territory size by conquest, or both together. They also raised the value of forming alliances outside of kin-based networks in order to subdue other neighboring groups.

Ten thousand years ago, the Neolithic revolution began to yield vastly larger amounts of food from cultivated crops and livestock, allowing rapid growth in human populations. But that advance did not change human nature. People simply increased their numbers as fast as the rich new resources allowed. As food again inevitably became the limiting factor, they obeyed the territorial imperative. Their descendants have never changed. At the present time, we are still fundamentally the same as our hunter-gatherer ancestors, but with more food and larger territories. Region by region, recent studies show, the populations have approached a limit set by the supply of food and water. And so it has always been for every tribe, except for the brief periods after new lands were discovered and its indigenous inhabitants displaced or killed.

The struggle to control vital resources continues globally, and it is growing worse. The problem arose because humanity failed to seize the great opportunity given it at the dawn of the Neolithic era. It might then have halted population growth below the constraining minimum limit. As a species we did the opposite, however. There was no way for us to foresee the consequences of our initial success. We simply took what was given us and continued to multiply and consume in blind obedience to instincts inherited from our humbler, more brutally constrained Paleolithic ancestors.

· 9 ·

The Breakout

TWO MILLION YEARS AGO the australopithecines of Africa, their genes spreading among multiple species, still roamed the savanna forests and grasslands of Africa. They walked on hind legs, which set them apart from all other primates that had ever existed. Their heads were ape-like in shape and dentition. Their brains were no larger than those of the great apes who lived around them. Their populations were scattered and small, and at any time all might have plunged to extinction. Within another half million years, all in fact were gone.

That is, all except one. The australopithecine radiation had yielded a single survivor, whose descendants were destined not only to persist but to dominate the world. At first, these ancestors of modern humanity were no more assured of a future than their close relatives had been. By two million years before the present, the favored australopithecine line had begun the transition to the still-larger-brained *Homo erectus*. This species had a brain smaller than that of present-day *Homo sapiens*, but it was able to shape crude stone tools and use controlled fire at campsites. Its populations spread out of Africa, blanketing the land up into northeastern Asia and pushing south all the way to Indonesia. *Homo erectus* was adaptable to an unprecedented degree for a primate. Some of its populations survived in the cold winters of present-day northern China, and others in the steaming tropical climate of Java. Across its great range, paleontologists have excavated fragments of every part of the *erectus* skeleton and repeatedly pieced them together. And in two sedimentary layers near northern Kenya's Lake Turkana, they discovered something as remarkable as skulls and thighbones: fossilized footprints. The impressions today have changed very little since a strolling *Homo erectus*, mud squishing between its toes, made them 1.5 million years ago.

Homo erectus, with a culture advanced well beyond that of its apish ancestors, and more adaptable to new and difficult environments, expanded its range to become the first cosmopolitan primate. It failed to reach only the isolated continents of Australia and the New World and the far-flung archipelagoes of the Pacific Ocean. Its great range buffered the species against early extinction. One of its genetic lines acquired potential immortality by evolving into *Homo sapiens*. The ancestral *Homo erectus* still lives. It is us.

At a far outlier of its range, *Homo erectus* produced a less fortunate offshoot. This was *Homo floresiensis*, a tiny, small-brained hominin that lived on Flores, a medium-sized island in the Lesser Sunda chain east of Java. Its fossil remains and stone tools date from 94,000 to only 13,000 years ago. At one meter in height and possessing a brain no larger than that of the African australopithecines, Flores man, also popularly known as the Hobbit, remains a tantalizing puzzle. It most likely originated as an extreme variant of *Homo erectus*, diverging during its isolation from the main Indonesian *erectus* populations. Its small size fits a loose rule of island biogeography: animal species isolated on islands and weighing less than twenty kilograms tend to evolve into relative giants (an example is the immense tortoises of the Galápagos), while those more than twenty kilograms tend to evolve into midgets (the dwarf deer of the Florida Keys). If its currently recognized status as a distinct hominin is correct, *Homo floresiensis* tells us a great deal about the vagaries of the evolutionary maze through which *Homo erectus* traveled to arrive at our own species. Its relatively recent extinction, following a long life, opens the possibility that it was erased, like our other sister species the Neanderthals, during the spread of all-conquering *Homo sapiens* around the world.

Homo sapiens, the successful descendant of *Homo erectus*, when viewed dispassionately is actually even more bizarre than the pygmy of Flores. Besides the bulging forehead, oversize brain, and long, tapering fingers, our species bears other striking biological features of the kind biological taxonomists call "diagnostic." This means that in combination, some of our traits are unique among all animals:

- A productive language based on infinite permutations of arbitrarily invented words and symbols.
- Music, comprising a wide array of sounds, also in infinite permutations and played in individually chosen mood-creating patterns; but, most definitively, with a beat.
- Prolonged childhood, allowing extended learning periods under the guidance of adults.
- Anatomical concealment of female genitalia and the abandonment of advertisement of ovulation, both combined with continuous sexual activity. The latter promotes female-male bonding and biparental care, which are needed through the long period of helplessness in early childhood.
- Uniquely fast and substantial growth in the brain size during early development, increasing 3.3 times from birth to maturity.
- Relatively slender body form, small teeth, and weakened jaw muscles, indicative of an omnivorous diet.
- A digestive system specialized to eat foods that have been tenderized by cooking.

Approximately 700,000 years ago, populations of *Homo erectus* were evolving larger brains. By inference, they had acquired at least the rudiments of some of the diagnostic traits of *Homo sapiens* just cited. Yet in this early period skulls were still far from modern. Archaic *Homo erectus* possessed bulging brow ridges, more projecting faces, and less lateral expansion of the overall skull than were to be the case for modern *Homo sapiens*. By 200,000 years before the present, the African ancestors had come anatomically closer to contemporary humans. The populations also used more advanced stone tools and may have engaged in some form of burial practice. But their skulls were still relatively heavy in construction. Only around 60,000 years ago, when *Homo sapiens* broke out of Africa and began to spread around the world, did people acquire the complete skeletal dimensions of contemporary humanity.

The ancestors who achieved the breakout from Africa and conquered Earth were drawn from a diverse genetic mix. Throughout their evolutionary past, during hundreds of thousands of years, they had been hunter-gatherers. They lived in small bands, similar to present-day surviving bands composed of at least thirty and no more than a hundred or so individuals. These groups were sparsely distributed. Those closest to each other exchanged a small fraction of individuals each generation, most likely females. They diverged genetically enough that the entire ensemble of bands (the metapopulation, as biologists call such a collectivity) was far more variable than the indigenous humans destined to achieve the breakout.

That difference persists. It has long been known that Africans south of the Sahara are far more diverse genetically than native peoples in other parts of the world. The magnitude of this disparity became especially clear when in 2010 all of the protein-coding sequences of the genome were published for four Bushman hunter-gatherers (also known as the San or Khoisan) from different parts of the Kalahari, plus a Bantu from a neighboring agricultural tribe in southern Africa. Amazingly, despite the outward physical similarity among them, the four San proved to differ more from one another than an average European does from an average Asian.

It has not escaped the attention of human biologists and medical researchers that the genes of modern-day Africans are a treasure house for all humanity. They possess our species' greatest reservoir of genetic diversity, of which further study will shed new light on the heredity of the human body and mind. Perhaps the time has come, in light of this and other advances in human genetics, to adopt a new ethic of racial and hereditary variation, one that places value on the whole of diversity rather than on the differences composing the diversity. It would give proper measure to our species' genetic variation as an asset, prized for the adaptability it provides all of us during an increasingly uncertain future. Humanity is strengthened by a broad portfolio of genes that can generate new talents, additional resistance to diseases, and perhaps even new ways of seeing reality. For scientific as well as for moral reasons, we should learn to promote human biological diversity for its own sake instead of using it to justify prejudice and conflict.

The *Homo sapiens* populations that spread from Africa into the Middle East and beyond took long journeys of the kind routine for modern-day travelers. Generation upon generation, the bands slogged cautiously on foot into the strange lands that lay before them. The pattern they appeared to follow was to venture a few tens of miles, settle, increase in numbers, then divide into two or more bands, capable of moving on into new territory. Apparently the initial invaders pressed north in this manner along the Nile Valley to the Levant, then spread out north and east. Quite possibly the first pioneers into the corridor made up only one or a very few bands. Within a few thousand years their descendants became a net of loosely connected tribes cast up on nearly the whole of the Eurasian continent.

This scenario of slow initial advance by a very few followed by local population growth is supported by two lines of evidence assembled by independent groups of researchers during the past ten years. First is the great genetic diversity of present-day southern Africans, suggesting that only a small part of the whole African population participated in the breakout. Second, analyses and mathematical models made of the amount of genetic differences among living human populations suggest that the pioneers created a "serial founder effect," with a few individuals moving out from an older, established population, then in turn serving as the source for the next emigration beyond. Eventually came multiple such spearheads radiating in many directions, and the human population coalesced.

Scientists have pieced together data from geology, genetics, and paleontology in order to envision more precisely how the out-of-Africa pattern began. Between 135,000 and 90,000 years ago, a period of aridity gripped tropical Africa far more extreme than any that had been experienced for tens of millennia previously. The result was the forced retreat of early humanity to a much smaller range and its fall to a perilously low level in population. Death by starvation and tribal conflict, both of which were to become routine in later historical times, must have been widespread in prehistory. The size of the total *Homo sapiens* population on the African continent descended into the thousands, and for a long while the future conqueror species risked complete extinction.

Then, finally, the great drought eased, and from 90,000 to 70,000 years ago tropical forests and savanna slowly expanded back to their previous ranges. Human populations grew and spread with them. At the same time, other parts of the continent became more arid, and the Middle East as well. With intermediate levels of rainfall prevailing throughout most of Africa, an especially favorable window of opportunity opened for the demographic expansion of pioneer populations out of the continent altogether. In particular, the interval was long enough to maintain a corridor of continuous habitable terrain up the Nile to Sinai and beyond, bisecting the arid land and allowing a northward sweep of colonizing humans. A second possible route was eastward, across the Bab el Mandeb Strait onto the southern Arabian Peninsula.

There followed the penetration of *Homo sapiens* into Europe by no later than 42,000 years before the present. Anatomically modern humans spread up the Danube River, entering the heartland of its sister human species the Neanderthals (*Homo neanderthalensis*). The latter populations had evolved in much earlier times from archaic human stock. Although genetically close to *Homo sapiens*, they were a distinct biological species, which on contact only rarely interbred with *sapiens*. Perhaps because the Neanderthals depended more on big game, they were poorly equipped to compete with skilled warriors who subsisted not only on big game but also on a wider variety of other animal and plant products. By 30,000 years before the present, *Homo sapiens* had entirely replaced them. *Homo sapiens* also replaced another species related to the Neanderthals, the recently discovered "Denisovans" of southern Siberia, known from remains in Denisova Cave in the Altai Mountains.

The remainder of the routes followed by the growing human populations, as best can be deduced by fossil and genetic evidence, extended outward into Asia and along the Indian Ocean coastline around 60,000 years ago. The colonists entered the Indian subcontinent and then the Malay Peninsula, while somehow making it across the straits to the Andaman Islands, where ancient aboriginal populations still exist. They apparently failed to reach the Nicobar Islands close by—where the genetic makeup of current inhabitants suggests a more recent Asian origin, 15,000 years before the present. The earliest human traces found to date in Indonesia, from the Niah Cave of Borneo, are 45,000 years old. The oldest from Australia, unearthed at Lake Mungo, date to 46,000 years. New Guinea was likely settled somewhat earlier. Major changes in the fauna of Australia, probably owing to predation and the use of burn-offs of low vegetation to drive game, give evidence that the date of the Australian incursion was at least 50,000 years before the present. The native people of New Guinea and Australia are thus truly aboriginals—direct descendants of the first modern humans to arrive in the same land they occupy today.

The question of exactly when anatomically modern *Homo sapiens* arrived in the New World, with its catastrophic impact on the virgin fauna and flora, has gripped the attention of anthropologists for many years. Like a photographic image in very slow developing fluid, the picture seems finally to be coming into focus. From genetic and archaeological studies across Siberia and the Americas, it now appears that a single Siberian population reached the Bering land bridge no sooner than 30,000 years ago, and possibly as recently as 22,000 years. In this period, the continental ice sheets had pulled enough water from the oceans to expose the Bering

FIGURE 9-1. *The first colonists of a new continent. Early in the history of modern humanity (*Homo sapiens*), tribes began burial ceremonies, which were antecedents or accompaniments of primitive religious belief. This reconstruction is a burial by early Australian aboriginals at Mungo, southeastern Australia, at least forty thousand years ago. Red ocher powder is being poured on the body of the corpse. (© John Sibbick. From* The Complete World of Human Evolution, *by Chris Stringer and Peter Andrews [London: Thames & Hudson, 2005], p. 171.)*

Land Bridge, while at the same time blocking entry into present-day Alaska. Around 16,500 years before the present, the retreat of the ice sheets cleared the way south, and a full-scale invasion through Alaska began. By 15,000 years before the present, as revealed by archaeological discoveries in both North and South America, the colonization of the Americas was well under way. It appears likely that the first populations dispersed along the recently deglaciated Pacific coastline, along land still exposed by the incomplete withdrawal of the ice sheets but nowadays mostly underwater.

Approximately 3,000 years ago, the ancestors of the Polynesian people began colonizing the Pacific archipelagoes. Starting at Tonga and proceeding stepwise eastward with large canoes designed for long voyages, they reached, by AD 1200, the extreme reaches of Polynesia, a triangle formed by Hawaii, Easter Island, and New Zealand. With this achievement of the Polynesian voyagers, the human conquest of Earth was complete.

· 10 ·

The Creative Explosion

POSSESSING BRAINS GROWN capable of global conquest, populations of *Homo sapiens* had broken from the African continent and spread, generation upon generation in a relentless wave, across all of the Old World. Almost imperceptibly at first but accelerating the pace here and there, they created increasingly complex forms of culture. Then, suddenly by geological standards, came the greatest of all advances. At multiple locations in the Neolithic dawn, the hunter-gatherers invented agriculture and formed villages, accompanied by chiefdoms and paramount chiefdoms, and finally states and empires. Cultural evolution during this time was (to borrow a term from chemistry) autocatalytic: each advance made other advances more likely. By the early centuries of recorded history, innovations were spreading rapidly back and forth across continents in both the Old and the New Worlds. It was, however, in the heartland of the Eurasian supercontinent that the process rose to the climax that was to change the world.

Three hypotheses have been offered by anthropologists to explain the creative explosion of culture. The first is that a major and transformative genetic mutation appeared in the African *Homo sapiens* population at about the time of the breakout into Eurasia. This view gains credence from the existence of our sister species *Homo neanderthalensis* for a hundred thousand years in Europe and the Levant, up to its disappearance only thirty millennia ago, without any major advance in its primitive stone technology. The Neanderthals devised neither visual art nor personal ornamentation. Oddly, throughout this static history, they had a larger brain than *sapiens*, and they had the challenge of a vast, constantly shifting environment. Judging from their anatomy and DNA, they probably could speak and, if so, very likely had complex languages. They took care of their injured, regardless of age, which was probably necessary for clan survival, since virtually every adult suffered broken bones from the reliance on big-game hunting. Yet for thousands of generations nothing much happened in Neanderthal culture. On the other hand, something immensely important did happen in the Africa-derived *sapiens*.

It seems unlikely, however, that a single, mind-changing mutation was responsible. A more realistic view is that the creative explosion was not a single genetic event but the culmination of a gradual process that began in an archaic form of *Homo sapiens* as far back as 160,000 years. This view has been supported by recent discoveries of the use of pigment that old, as well as personal ornaments and abstract design scratched on bone and with ocher dating from between 100,000 and 70,000 years ago.

The third hypothesis advanced by anthropologists is that cultural innovation and its adoption rose and fell with the severe changes occurring during the same period in climate, which had dire effects on human population size and growth. Some of the innovations disappeared only to be reinvented later, while others caught fire and held thereafter until the breakout period. This view is supported by the earliest archaeological record suggesting that African artifacts, including shell beads, bone tools, abstract engravings, and the improved shaping of stone projectile points, were followed by their apparent widespread disappearance during a long and especially intense climatic deterioration between 70,000 and 60,000 years ago. The discontinuity was followed in turn by their reappearance around 60,000 years ago, at approximately the time of the breakout. It is believed that during the period of climatic deterioration, populations declined and became more scattered, disrupting social networks and causing a loss of some cultural practices. When the climate improved, and populations grew and expanded again, the innovations were reinvented and others added to them in time to be carried out of Africa during the global colonization. Just as in modern culture (albeit for different reasons), innovations winked on and off, with a few taking hold and spreading.

In fact, the three hypotheses are not mutually exclusive. They can be fitted together in a single scenario. Genetic evolution was certainly occurring during the entire span of time from the breakout to the spread of population across the Old World. According to one study, the rate of origin of new genetic mutations was relatively low and steady until about 50,000 years ago, then rose to a peak approximately 10,000 years ago, at the start of the Neolithic revolution. During the same period, human population growth also accelerated. As a consequence, more genetic mutations occurred, and in addition, by the sheer increase in numbers of people, more cultural innovations were achieved.

When geneticists compared the genomes (entire genetic codes) of modern chimpanzees and humans as a yardstick, they deduced that about 10 percent of the amino acid changes since the divergence of the two species from common stock six million years ago have been adaptive—in other words, guided by natural selection that favored their survival across generations. A variety of other studies have confirmed that during the breakout and spread, evolution was actually occurring. Overall, body size decreased a small amount, while brain size and teeth grew proportionately smaller. Other traits evolved in the outlier populations of Europe and Asia, and then in the Americas. Such a pattern is entirely to be expected. Abundant variation within and between populations on which natural selection could act became available. Differences also arose from random sampling during advances of the populations, causing "genetic drift" independent of adaptation. (To visualize genetic drift, a product of chance, consider flipping a coin, then doubling it if it turns up heads and throwing it away if it turns up tails. Essentially this process determines the fate of a mutated gene, unless it is either favorable or unfavorable for the organisms carrying it.) The most likely cause of such genetic drift was the founder effect, owing to chance differences between bands belonging to the same community of bands during the spread of populations. When a first group departed in one direction during its emigration and a second group stayed or traveled in another direction, each carried its own distinct collective set of genes, since each was only a fraction of the whole existing in the mother population. As a result, skin color, height, percentages of blood types, and other nonvital hereditary traits shifted a bit in one direction or another over distances as short as a few hundreds of kilometers.

Mutations are random changes in the DNA. They can occur by a simple change in a single letter (that is, in a base pair, AT to GC or the reverse), by multiplication of an existing letter (for example, AT to ATATAT), or by the moving of letters to new locations on the same chromosome or a different chromosome. Each gene typically consists of thousands of such letters. They are also highly variable in this number. For example, 23 genes per million base pairs are on the human chromosome 19, but only 5 genes per million base pairs are on chromosome 13.

When the burst of new mutations inevitably occurred following the breakout from Africa because of the vast overall increase in population size, humans passed through two phases of evolution. In the first period, all mutations were at very low levels, since under all conditions they typically arise at rates less than one in ten thousand individuals and as low as one in billions. While still at such minimal, "mutational" levels, most of the changes disappear, either because they reduce the fitness of the individuals who carry them or by simple chance (genetic drift), or some combination of the two. If, however, the new mutant gene reaches a frequency of about 30 percent, it is likely to increase still further. Eventually, during the second phase of evolution, the mutant form of the gene (mutant allele) may completely replace the older, competing form of the same gene (older allele). Another possibility is that the combination of the two alleles in the same person (who is then called a heterozygote for that gene) does better than does either one of the alleles in double dose (homozygotes). In that case, the frequency of the mutant will reach equilibrium with the old gene at less than complete fixation of either one. The textbook example is sickle-cell anemia, the gene for which occurs throughout malarial areas from Africa to India. Two sickle-cell genes give you severe anemia, with a high risk of death. Two normal genes leave you at high risk of contracting malaria. One sickle-cell gene and one normal gene together (the heterozygote condition) protect you from both. The result is a high frequency of both genes in the malarial areas, kept more or less in equilibrium by the selection pressure of malaria.

Since the split of the human line from that of the chimpanzees, the human line has followed a pattern apparently consistent with that of animals in general. Its existence, if proven, has profound significance for understanding how the human condition was attained. The pattern is that coding genes, which control changes in the structure of enzymes and other proteins, dominate expression of traits in particular tissues, such as those affecting the immune response, the sense of smell, and sperm production. In contrast, noncoding genes, which regulate hereditary developmental processes prescribed by coding genes, are more active in the development and function of the nervous system. Although the analyses on which this distinction is based are preliminary, it is considered probable that noncoding changes have been of key importance in the evolution of cognition, in other words, the changes that made us human.

Which traits of cognition in fact have evolved through mutations and natural selection, both coding or noncoding? Very likely, all of them. Twin studies, in which difference between identical twins (who are genetically identical, because of their origin in a single fertilized egg) is compared with difference between fraternal twins (born of separately fertilized eggs, hence genetically as different as between siblings born at different times), suggest that personality traits such as introversion-extroversion, shyness, and excitability are subject to strong genetic influences. The amount of variation due to differences in genes in a given population usually falls between one-fourth and three-fourths.

Of at least equal importance in the evolutionary origin of advanced social behavior, of humans or any other kind of organism, is the genetic influence on variation of social networks. We would expect there to be some amount of such genetic control, in accordance with Turkheimer's "first law" of behavior genetics—that all traits vary to some extent among people because of differences in genes. (The other two "laws" are "The effect of being raised in the same family is smaller than the affect of genes" and "A substantial portion of the variation in complex human behavioral traits is not accounted for by the effects of genes on families.") Interactions, in particular, have so many sources in individual behavior, each of which is likely to show genetic variation, that it would be a major surprise if their combinations were found to add up to none at all in social networks. In fact, personal networks are highly variable in size and strength, and heredity plays a role. A recent study has found that variation in the number of people one person has in contacts or in social ties, as well as variation in transitivity—the likelihood that any two of a person's contacts are connected to each other's contacts—are both about half due to heredity. On the other hand, the number of other group members whom individuals view as friends is not genetically influenced, at least not within ordinary statistical limits of the measures taken.

Taking into account the genetic and archaeological evidence available to date, and now growing rapidly, I believe the long-term trajectory leading up to the breakout and afterward can be plotted roughly as follows. In attempting it, I think it useful first to mention an analogy from biogeography and ecology. Cultural innovations can be compared to species of organisms that accumulate during the buildup of numbers of species colonizing an ecosystem, such as a newly formed pond, copse, or small island. There is a turnover in culture traits in a band of humans, just as there is in species that colonize an ecosystem. Some cultural innovations persisted in the

African bands following their spread. Others, as the archaeological evidence on body ornaments and projectile points shows, passed into extinction, usually to be reintroduced later either by invention or else by contact with other bands. At first, the human bands on the African continent were small and isolated. Their numbers and the average size of each waxed and waned in the face of changes in the climate and the availability of habitable terrain. As the environment became more favorable before and during the breakout from Africa, the numbers of bands and their population size increased. As a consequence, the rate at which they acquired innovations also increased.

During this critical period of human prehistory, 60,000 to 50,000 years ago, the growth of cultures became autocatalytic. At first, as I have suggested, the growth was slow, then faster and still faster and yet again faster, in the manner of chemical and biological autocatalysis. The reason is that the adoption of any one innovation made adoption of certain others possible, which then, if useful, were more likely to spread. Bands and communities of bands with better combinations of cultural innovations became more productive and better equipped for competition and war. Their rivals either copied them or else were displaced and their territories taken. Thus group selection drove the evolution of culture.

In a very early time, from the Late Paleolithic period through the Mesolithic period, the cultural evolution of humanity ground forward slowly. At the beginning of the Neolithic period, 10,000 years before the present, with the invention of agriculture and villages and food surpluses, cultural evolution accelerated steeply. Then, thanks to the expansion of trade and by force of arms, cultural innovations not only increased faster but also spread much farther. There was still a turnover in innovations, but now, given the sheer mass of people and tribes making them, some were original and powerful enough to be overwhelming in their impact. Such revolutionary advances as writing, astronomical navigation, and guns were at first rare, imperfect, and fragile. Some disappeared, only to

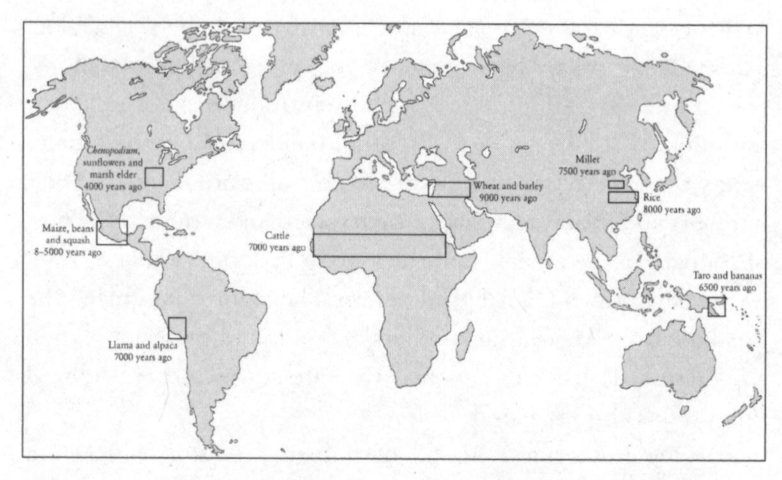

FIGURE 10-1. *The centers of the eight known independent origins of agriculture, including animal husbandry, and the approximate dates they occurred. (From Steven Mithen, "Did farming arise from a misapplication of social intelligence?"* Philosophical Transactions of the Royal Society ***B*** *362: 705–718 [2007].)*

reappear later. Like sparks from a fire, each had a chance to catch, burst into flame, and spread.

Archaeologists have described some of the key mental concepts that thus took hold and spread during 10,000 and 7,000 years before the present.

- The mastery of stone was completed, taking toolmaking far from the simple knapping of available stones used in the Mesolithic to a far more sophisticated procedure. Axes and adzes invented in the Neolithic were made by a series of steps. Each blade was first flaked out to the right shape from a block of fine-grained rock. Then it was shaped more finely by chipping out progressively smaller flakes. Lastly, rough spots on the surface were removed by precise chisel work or grinding. The final product was a blade with a smooth surface, sharp-edged, and flattened or rounded to the form needed.

- Neolithic toolmakers invented the concept of a hollow structure, with an outer and an inner surface. Accordingly, they devised containers of useful shapes variously out of wood, leather, stone, or clay.

- The toolmakers also figured out how to reverse the steps of their ancient manufacture, by starting with small objects and assembling them into larger ones. By this means weaving was invented, and increasingly more elaborate and spacious dwellings were erected.
- A pivotal change—ultimately important not only for humanity but also for the rest of life—was the new conceptions of the environment formed in the minds of the fledgling farmers and villagers. Natural habitats were no longer wild places in which to hunt and gather food, and occasionally burn over with ground fires. The habitats instead became land to be cleared for agriculture. This particular conception, that wildland is something to be replaced, has been a mental fixation of most of the world's population to this day.

The roots of agriculture go back to the breakout period or slightly afterward, at least 45,000 years ago, when fire was used to drive and capture game. At that time, at least some of the human bands must have recognized, as Australian aboriginals do today, that ground fires are followed in savannas and dry forest by the growth of increased amounts of fresh, edible vegetation. Nutritious underground tubers are for a while also easier to find and excavate. As revealed by recent detailed studies of native Mexican crops, the next step was made possible by the establishment of long-term human settlements. The inhabitants of Mexico and other parts of Mesoamerica began to cultivate productive trees and other plants, such as agave, opuntia, gourds, and the leguminous tree *Leucaena*, simply by allowing them to grow to the exclusion of other plants around their dwellings. (It is interesting that a few species of ants do the same thing.) The next step was equally serendipitous. Some of these earliest garden species accidentally hybridized with other, similar species, or else multiplied their number of chromosomes, or made both such alterations together, producing new strains that were even more valuable as food. When they appeared and were sampled by harvesters, they were selected over others. Thus began tree domestication by artificial selection, and the practice of plant breeding. About the same time, or even before, domestication was practiced on animals captured in the wild and converted into pets and livestock. From 9,000 to 4,000 years ago, the trend was furthered to include many new strains of plants and animals in at least eight major centers in the Old and New Worlds. Agriculture was thus launched as the primary human occupation.

The past ten millennia have been a period of extraordinary change for both *Homo sapiens* and the rest of the biosphere. Cultural evolution is still accelerating, and that raises a fundamental question: are we also evolving genetically? Medical research, added to a deepening analysis of the three billion nucleotide letters of the human genome, has revealed that evolution is indeed still occurring in human populations. Because of the emphasis on medicine in human genetics, the great majority of genes identified thus far as subject to natural selection are those that provide resistance to disease. The list is growing of mutations that appeared and spread in recent millennia: CGPD, CD406, and the sickle-cell gene, each providing some degree of natural protection against malaria; CCR5 against smallpox; AGT and CY3PA against hypertension; and ADH against parasites sensitive to aldehydes. There are also genetic mutations of recent origin that affect physiological traits, including the classic case of the adult lactose-tolerant gene that permits consumption of milk and milk products. The highland Tibetans, living with low levels of oxygen, have acquired EPAS1, which prescribes increased production of hemoglobin, the key to performance at high altitudes. From all that we know of its fundamental processes, evolution in the human species has been in recent times and will continue to be inevitable.

Human geneticists agree that most geographical variants in anatomy and physiology, sufficiently restricted to one geographic area to be popularly classified as racial, are due not to localized natural selection but to emigration of different genetic types and random fluctuations in local frequencies of genes leading to genetic drift. Exceptions include skin color, the geographic variation of which is attributed to protection from ultraviolet radiation in sunlight, which increases toward the equator. They also include the unusually broad faces of Greenland Eskimos and the Buriat people of Siberia, a feature that minimizes the surface area as a protection against extreme cold.

Changes in gene frequency due to evolution at the level of one gene or a small ensemble of genes whether linked on the same chromosome or not, and referred to by biologists as microevolution, are expected to continue as a natural process into the indefinite future. For the immediate future, however, emigration and ethnic intermarriage have taken over as the overwhelmingly dominant forces of microevolution, by homogenizing the global distribution of genes. The impact on humanity as a whole, even while still in this current early stage, is an unprecedented dramatic increase in the genetic variation within local populations around the world. The increase is matched by a reduction in differences *between* populations. Theoretically, if the flow continues long enough, the population of Stockholm could come to be the same genetically as that in Chicago or Lagos. Overall, more kinds of genotypes are being produced everywhere. This change, unique in human evolutionary history, offers a prospect of an immense increase in different kinds of people worldwide, and thereby newly created physical beauty and artistic and intellectual genius.

The geographical homogenization of *Homo sapiens* appears unstoppable, but it will in time be overlaid by yet another, presumably final force of evolution, volitional selection. Engineering by gene substitution in embryos will soon be a reality at the experimental level, and thereafter be used to combat hereditary disease. In time, it will become a routine therapeutic procedure in medical practice. Soon afterward, depending upon the outcome of a whole new level of moral debate certain to be intense, the genetic makeover of normal children in the embryo stage might (or might not) become a major branch of the biomedical industry. I hope, and am inclined to believe on moral grounds, that this form of eugenic manipulation will never be permitted, in order that humanity can at the very least avoid the socially corrosive effects of nepotism and privilege it is bound to serve.

I am further inclined to discount the widespread belief that robotic intelligence will in the near future overtake and potentially replace human intelligence. This will certainly occur in the categories of raw memory, computation, and synthesis of information. Algorithms might in time be written that simulate emotional responses and human-like processes of decision-making. Yet even at their most extreme and effective, these creations will still be robots. If anything can be drawn from the picture of human condition assembled by science, it is that as a result of prehistory our species is extremely idiosyncratic in both emotion and thought. Our particular passage through the evolutionary maze stamped our DNA at every major step along the way. Humanity is indeed unique, perhaps more than we ever dreamed. Yet despite our singularity on this planet at this time, we are psychically only one of a large number of species of roughly humanoid grade or above that might have occurred or, should we extinguish ourselves, might yet occur in the billions of years left to the biosphere.

Scientists have only begun to probe the neural pathways and endocrine regulation of the subconscious that impose a decisive influence on feeling, thought, and choice. Further, the mind consists not just of this inner world but also of the sensations and messages that flow in and out of it from all other parts of the body. To advance from robot to human would be a task of immense technological difficulty. But why should we even wish to try? Even after our machines far exceed our outer mental capacities, they will not have anything resembling human minds. In any case, we do not need such robots, and we will not want them. The biological human mind is *our* province. With all its quirks, irrationality, and risky productions, and all its conflict and inefficiency, the biological mind is the essence and the very meaning of the human condition.

· 11 ·

The Sprint to Civilization

ANTHROPOLOGISTS RECOGNIZE THREE levels of complexity among human societies. At the simplest level, hunter-gatherer bands and small agricultural villages are by and large egalitarian. Leadership status is granted individuals on the basis of intelligence and bravery, and through their aging and death it is passed to others, whether close kin or not. Important decisions in egalitarian societies are made during communal feasts, festivals, and religious celebrations. Such is the practice of the few surviving hunter-gatherer bands, scattered in remote areas, mostly in South America, Africa, and Australia, and closest in organization to those prevailing over thousands of years prior to the Neolithic era.

Chiefdoms, the next level of complexity, also called rank societies, are ruled by an elite stratum who upon debility or death are replaced by members of their family or at least those of equivalent hereditary rank. That was the dominant form of societies around the world at the beginning of recorded history. Chiefs or "big men" rule by prestige, largesse, the support of elite members below them—and retribution against those who oppose them. They live on the surplus accumulated by the tribe, employing it to tighten control upon the tribe, to regulate trade, and to wage war with neighbors. Chiefs exercise authority only on the people immediately around them or in nearby villages, with whom they interact as needed on a daily basis. In practice this means subjects who can be reached within half a day traveling by foot. The reach is thus a maximum of twenty-five to thirty miles. It is to the advantage of chiefs to micromanage the affairs of their domain, delegating as little authority as possible in order to reduce the chance of insurrection or fission. Common tactics include the suppression of underlings and the fomenting of fear of rival chiefdoms.

States, the final step up in the cultural evolution of societies, have a centralized authority. Rulers exercise their authority in and around the capital, but also over villages, provinces, and other subordinate domains beyond the distance of a one day's walk, hence beyond immediate communication with the rulers. The domain is too far-flung, the social order and communication system holding it together too complex, for any one person to monitor and control. Local power is therefore delegated to viceroys, princes, governors, and other chief-like rulers of the second rank. The state is also bureaucratic. Responsibility is divided among specialists, including soldiers, builders, clerks, and priests. With enough population and wealth, the public services of art, sciences, and education can be added—first for the benefit of the elite and then, trickling down, for the general public. The heads of state sit upon a throne, real or virtual. They ally themselves with the high priests, and clothe their authority with rituals of allegiance to the gods.

The ascent to civilization, from egalitarian band and village to chiefdom to state, has occurred through cultural evolution, not through changes in genes. It is a spring-loaded change, unfolding in a manner parallel to, but far grander than, the one propelling insect groups from aggregates to families, then to eusocial colonies with their castes and division of labor.

The prevailing theory among anthropologists is that whenever tribes can acquire more territory through aggression or technology they do so, and thereby acquire more resources. They may then continue to expand if they are able, ultimately blossoming into empires or fissioning into new, competing states. With larger size and farther reach comes greater complexity. And as with complexity of any physical or biological system, the society, in order to achieve stability and survive and not quickly crumble, must add hierarchical control. A state-level hierarchy is a system composed of interacting subsystems, all together hierarchical in structure, descending in sequence until the lowest level of subsystem is reached, in this case the individual citizen of the state. A true system is "decomposable" into subsystems (such as infantry companies and municipal governments) that interact with one another. Individuals in one subsystem do not have to interact with individuals in other subsystems at the same level. A system that is highly decomposable in this manner is likely to work better than one that is not. "On theoretical grounds," the mathematical theorist Herbert A. Simon said in his pioneering paper on the subject, "we could expect complex systems to be hierarchies in a world in which complexity had to evolve from simplicity. In their dynamics, hierarchies have a property, near-decomposability, that greatly simplifies their behavior. Near-decomposability also simplifies the description of a complex system, and makes it easier to understand how the information needed for the development or reproduction of the system can be stored in reasonable compass."

Translated to the cultural evolution of simpler societies into states, Simon's principle suggests that hierarchies work better than unorganized assemblages and that they are easier for their rulers to understand and manage. Put another way, you cannot expect success if assembly-line workers vote at executive conferences or enlisted men plan military campaigns.

Why call the evolution of human societies into civilization cultural as opposed to genetic? There exist multiple lines of evidence to support this conclusion. Not least is the fact that infants of hunter-gatherer societies raised by adoptive families in technologically advanced societies mature as capable members of the latter—even though the ancestral lines of the child have been separated from those of its adoptive parents by as long as 45,000 years—in, for example, Australian aboriginal children raised by white families. That length of time has been enough to produce genetic differences between human populations through combinations of natural selection and genetic drift. But the known traits that were genetically changed are, as we have seen, primarily in resistance to disease and adaptation to local climates and food sources. No statistical genetic differences between entire populations have yet been discovered that affect the amygdala and other controlling circuit centers of emotional response. Nor is any genetic change known that prescribes average differences between populations in the deep cognitive processing of language and mathematical reasoning—although such may yet be detected.

The stereotypes by which inhabitants of different nations, cities, and villages are often characterized might also have some hereditary basis in fact. However, the evidence suggests that the differences have a historical and cultural origin rather than a genetic one. As such, whatever hereditary variation among cultures that does exist is dwarfed when put in a genetic evolutionary time scale. Italians may be more voluble on average, Englishmen more reserved, Japanese more polite, and so on, but the average between populations of such personality traits are hugely outweighed by their variation within each population. It turns out, remarkably, that the variation is closely similar from one population to the next. Such was the observation of the American psychologist Richard W. Robins during his residence in a remote village of the West African nation of Burkina Faso.

> While there, I was struck by the degree to which everyone seemed so different yet so familiar at the same time. Despite dramatic differences in cultural customs and practices, the Burkinabe people seemed to fall in love, hate their neighbors, and care for their children in much the same way, and for many of the same reasons, as people in other parts of the world. Indeed, there is a core to human mentality and social behavior that cuts across nations, cultures, and ethnic groups. Even such profoundly different countries as Burkina Faso and the United States do not differ substantially in the average personality tendencies of their people. . . .
>
> Against this backdrop of human universals, it is quite clear that

individual variability exists: Some Burkinabe (or Americans) are shy and others sociable, some friendly and others disagreeable, and some driven to attain high status in their community while others lack the same drive.

Of the very large array of personality traits researched by psychologists, most can be divided into five broad domains: extroversion versus introversion, antagonism versus agreeableness, conscientiousness, neuroticism, and openness to experience. Within populations each of these domains contains substantial heritability, mostly falling between one-third to two-thirds. This means that of the total variation of scores in each domain—the fraction due to differences in genes among individuals—falls somewhere between one-third and two-thirds. So from inheritance alone we would expect to find substantial variation in a population such as that in the Burkina Faso village. Added to differences in experience from one person to the next, especially during the formative periods of childhood, we should expect to find even greater variation, but more or less consistently from village to village, and from country to country.

Does such substantial variation exist universally, and is it the same from one population to the next, or different? The variation turns out to be consistently great and universally to the same degree across populations. Such was the result of an extraordinary study conducted by a team of eighty-seven researchers and published in 2005. The degree of variation in personality scores was similar across all of forty-nine cultures measured. The central tendencies of the five domains of personality differed only slightly from one to the next, in a way that was not consistent with prevailing stereotypes held by those outside the cultures.

A reason for doubting that large-scale genetic differences exist is the nearly simultaneous origin of state-based civilizations in the six best-analyzed locations around the world, when compared with the relatively enormous geological span of evolutionary changes in human anatomy. Each one in turn followed relatively soon upon the domestication of crops and livestock, although in other parts of the world these innovations had not yet yielded state-level societies. In Egypt, the earliest primary state (that is, the earliest among those independently evolved) was at Hierakonpolis, between Upper Egypt and Lower Nubia, at 3400–3200 BC. In the Indus Valley of Pakistan and northwestern India, mature Harappan settlements had evolved into a state by 2900 BC. And in China the earliest primary state appears to have been at Erlitou, beginning in 1800–1500 BC. Finally, the first documented rise of a primary state in the New World is that in Mexico's Oaxaca Valley, between 100 BC and AD 200. The arid north coast of Peru was the site of the independently evolved Moche State, which began during AD 200–400.

It is highly unlikely that primary states emerged around the world as the result of convergent genetic evolution. It is all but certain that they appeared autonomously as elaborations of already existing genetic predispositions shared by human populations through com-

TABLE 11-1. The origin of the earliest known independently evolved state in the New World, based upon archaeological evidence from the Oaxaca Valley of Mexico. [Modified from Charles S. Spencer, "Territorial expansion and primary state formation," *Proceedings of the National Academy of Sciences, U.S.A.* 107(16): 7119–7126 (2010).]

	Tiers in settlement hierarchy	*Palace*	*Multi-room temple*	*Long-distance conquests*	*Valley-wide integration*
AD 200 – 100 BC	4	Yes	Yes	Yes	Yes
100 BC – 300 BC	4	Yes	Yes	Yes	No
300 BC – 500 BC	3	No	No	No	No
500 BC – 700 BC	3	No	No	No	No

mon ancestry and dating back to the breakout period some 60,000 years ago. Their explanation is supported by the relatively swift rise of a primary state on the Hawaiian island of Maui. Prehistoric settlers apparently reached this island around AD 1400 with agricultural capability. By AD 1600 the population has expanded significantly, temples were built, and a single ruler took control of two formerly independent villages. The rate of change was faster than that in the Oaxaca Valley, where 1,300 years passed from the first known village to the construction of the first state temple.

By the time of the breakout, African populations were engraving ostrich eggshell containers. Even earlier (100,000 to 70,000 BP), they had been using pieces of red ocher, pierced shell beads, and advanced tools. These artifacts, the oldest of which date halfway back to the origin of anatomically modern *Homo sapiens* itself, are as sophisticated as some of those fashioned by modern hunter-gatherers.

The rudiments of civilization also arrived close behind the dawn of agriculture, or even before it. At Göbekli Tepe, an isolated site in Turkey on the Euphrates River, archaeologists have excavated a hilltop temple about 11,000 years in age. There are pillars and stone slabs, many of which are carved with the images of familiar animals—mostly crocodiles, boars, lions, and vultures, and one scorpion. There are other, unknown but fierce-looking creatures whose visages may have been inspired by nightmares or drug-induced delusions. Some researchers at Göbekli Tepe have concluded that because no remains of nearby villages have been found, the monuments are the work of nomadic hunter-gatherers who assembled there occasionally for religious ceremonies. Others, however, believe that such villages, large enough to have supported many workers, will in time be found.

There is a rule that applies to both archaeology and paleontology: *no matter how old the earliest known fossil or evidence of a human activity is, there is always somewhere and remaining to be discovered evidence of something at least a bit older.* The principle has been well borne out in the case of literacy. The earliest known writing is that of the Mesopotamian culture of Sumer and the early Egyptian culture, dating to 6,400 years before the present, hence more than halfway back to the beginning of the Neolithic era. That is followed by the first known script of the Indus Valley culture in present-day Pakistan (4,500 years BP), the Shang Dynasty of China (3,500–3,200 BP), and the Olmecs of Mesoamerica (2,900 BP). All of these ancient scripts, however, present a daunting mystery. It is rarely clear to what extent the various cuneiform symbols and pictographs represent abstractions as opposed to real entities, and whether they denote syllables and sounds of the language or, alternatively, concepts designated by unknown words used in a now vanished speech. No scholar doubts, however, that once perfected the written records they created gave an enormous advantage to their inventors.

If the shift from chiefdoms to states has been spring-loaded and cultural, how are we to account for the disparities in present-day societies? The differences are enormous. If countries are ranked by their per capita incomes, those in the top 10 percent are on average approximately 30 times richer than those in the bottom 10 percent, while the richest are 100 times richer than the poorest. The consequences of such variation on the quality of life are staggering. In the poorest countries live more than one billion people, some 15 percent of the world population, existing in what the United Nations classifies as absolute poverty. They lack adequate housing, sanitation, clean water, health care, education, and dependable food. The inhabitants of richer nations, some close by the poorest, enjoy all these benefits, including air travel and vacations. According to Jared Diamond in his celebrated 1997 work *Guns, Germs, and Steel: The Future of Human Societies*, and substantiated in analyses by the Swedish economists Douglas A. Hibbs Jr., Ola Olsson, and others, a persuasive answer can be found in geography. Just before the origins of agriculture some 10,000 years ago, a combination of conditions gave peoples of the Eurasian supercontinent an enormous opportunity to further the cultural revolution soon to be made possible. The great size of the continent, its vast breadth east to west, and its augmentation by the biologically rich lands of the Mediterranean perimeter resulted in an endowment of more plant and animal species locally suitable for domestication than existed on islands and other continents. Knowledge of crops and farm animals and the technology to build and store surpluses were more quickly spread from village to village, and then across the widening territories of the early states. The size and fruitfulness of this Eurasian heartland, not the emergence of a human genome endemic to any particular place, led to the Neolithic revolution.

III

How Social Insects Conquered the Invertebrate World

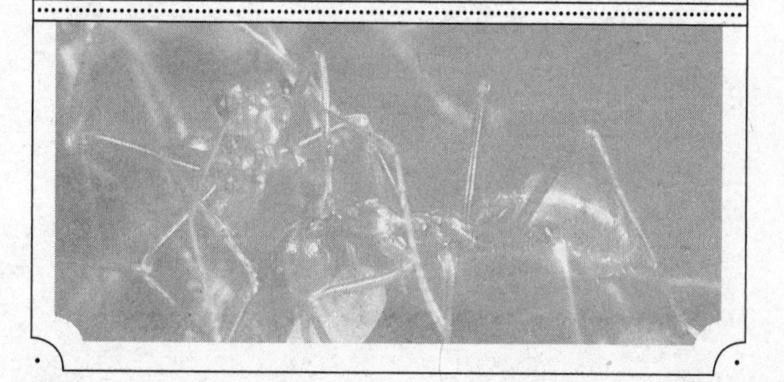

· 12 ·

The Invention of Eusociality

THE KEY TO THE ORIGIN of the human condition is not to be found in our species exclusively, because the story did not start and end with humanity. The key is to be found in the evolution of social life in animals as a whole.

When you look at the full panorama of social behavior in the animal kingdom, and not just the part of it represented by human beings, a pattern stands out sharply. Seldom considered by evolutionary biologists in the past, it comprises two phenomena connected by cause and effect. The first phenomenon is that animals of the land environment are dominated by species with the most complex social systems. The second phenomenon is that these species have evolved only rarely in evolution. They have arisen through many preliminary steps across millions of years of evolution. Humanity is one of the animal species.

The most complex systems are those possessing eusociality—literally "true social condition." Members of a eusocial animal group, such as a colony of ants, belong to multiple generations. They divide labor in what outwardly at least appears to be an altruistic manner. Some take labor roles that shorten their life spans or reduce the number of their personal offspring, or both. Their sacrifice allows others who fill reproductive roles to live longer and produce proportionately more offspring.

The sacrifices within the advanced societies go far beyond those between parents and their offspring. They extend to collateral relatives, including siblings, nieces, and nephews, and cousins at various degrees of remove. Sometimes they are bestowed on genetically unrelated individuals.

A eusocial colony has marked advantages over solitary individuals competing for the same niche. Some of the colony members can search for food while others protect the nest from enemies. A solitary competitor belonging to another species can either hunt for food or defend its nest, but not do both at the same time. The colony can send out multiple foragers and stay home all at the same time, forming a webwork of surveillance both within and around the nest. When food is found by one colony member, it can inform the others, who then converge on the site like a closing net. When assembled, the nestmates have the ability to fight as a group against rivals and enemies. They can transport large quantities of food more rapidly to the nest, before competitors arrive. With multiple individuals serving as construction workers, the nest can quickly be made larger, its structure architecturally more efficient, and its entrances more easily defended. The nest can also be climate-controlled to some extent. The nests of the mound-building termites of Africa and leaf-cutting ants of the Americas represent the ultimate state: they are designed to be air-conditioned, freshening and circulating air without further action on the part of the inhabitants.

Large colonies of some species can also apply military-like formations and mass attacks to overcome prey that are invulnerable to solitary individuals. The driver (or army) ants of Africa are among the ultimates in this adaptation. They march in columns of up to millions, consuming most small animals in their path. The hordes of these and other army ant species are also unique among insects in their ability to defeat and consume large colonies of termites, wasps, and other kinds of ants.

The twenty thousand known species of eusocial insects, mostly ants, bees, wasps, and termites, account for only 2 percent of the approximately one million known species of insects. Yet this tiny

FIGURE 12-1. *The two conquerors of Earth. Social insects rule the insect world. A single colony of African driver ants, one of which is depicted here on a foraging expedition, contains as many as 20 million workers. (From Edward O. Wilson,* Success and Dominance in Ecosystems: The Case of the Social Insects *[Oldendorf/Luhe, Germany: Ecology Institute, 1990].)*

minority of species dominate the rest of the insects in their numbers, their weight, and their impact on the environment. As humans are to vertebrate animals, the eusocial insects are to the far vaster world of invertebrate animals. Among creatures larger than microorganisms and roundworms, eusocial insects are the little things that run the terrestrial world.

Weaver ants are among the most abundant insects in the canopies of tropical forests, from Africa to Asia and Australia. They form chains of their own bodies in order to pull leaves and twigs together to create the walls of shelters. Others weave silk drawn from the spinnerets of their larvae to hold the walls in place. That done, they cover

FIGURE 12-2. *At one typical Amazonian locality, ants were found to outweigh all the vertebrate animals (represented here by a jaguar) four to one. (From Edward O. Wilson,* Success and Dominance in Ecosystems: The Case of the Social Insects *[Oldendorf/Luhe, Germany; Ecology Institute, 1990]. Based on E. J. Fittkau and H. Klinge, "On biomass and trophic structure of the central Amazonian rain forest ecosystem,"* Biotropica *5[1]: 2–14 [1973].)*

the football-sized shelters with silken sheets. Occupying hundreds of these aerial pavilions, a single colony of weaver ants composed of the mother queen and hundreds of thousands of her daughter workers can dominate several trees at a time.

From Louisiana to Argentina, immense colonies of leafcutter ants, the most complex social creatures other than humans, build cities and practice agriculture. The workers cut fragments from leaves, flowers, and twigs, carry them to their nests, and chew the material into a mulch, which they fertilize with their own feces. On this rich material, they grow their principal food, a fungus belonging to a species found nowhere else in nature. Their gardening is organized as an assembly line, with the material passed from one specialized caste to the next all the way from the cutting of raw vegetation to the harvesting and distribution of the fungus.

In one Amazon site, two German researchers accomplished the prodigious task of weighing all of the animals in a single hectare of rainforest. They found that ants and termites together compose almost two-thirds of the weight of all the insects. Eusocial bees and

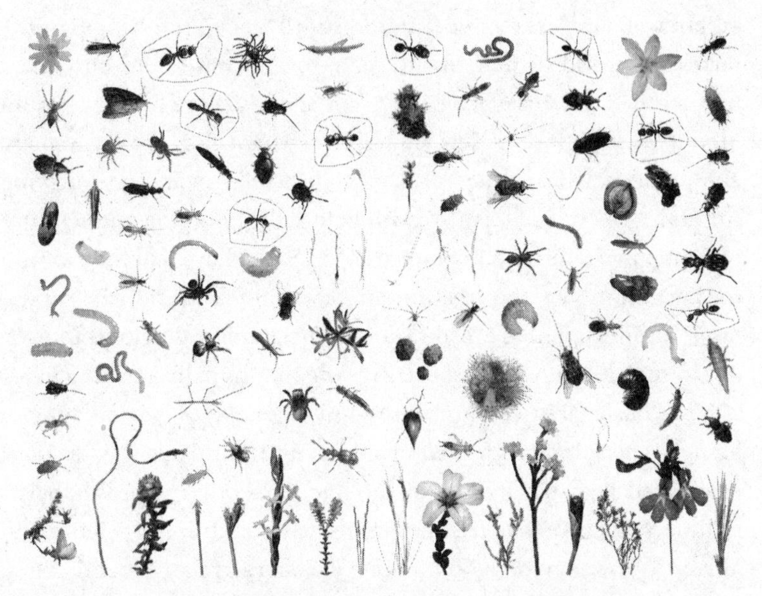

FIGURE 12-3. *The ubiquity of ants. Laid out here is the variety of small organisms found in one cubic foot of soil and leaf litter on a limb of a strangler fig at Monteverde, Costa Rica. Eight of the one hundred individuals present were ants (encircled). (From Edward O. Wilson, "One cubic foot," David Liittschwager* National Geographic, *February 2010, pp. 62–83. Photographs by David Liittschwager. David Liittschwager / National Geographic Stock)*

wasps added another tenth. Ants alone weighed four times more than all the terrestrial vertebrates—that is, mammals, birds, reptiles, and amphibians combined. Other researchers determined that ants alone make up two-thirds of the insects in the high canopy of another Amazonian locality.

Ants are not quite a thick layer of insect tissue upon the earth. They are much sparser in the cold conifer forests in both the Northern and the Southern Hemispheres, and they peter out just north of the Arctic Circle and near the tree line on tropical mountains. There are no ants as well on Iceland, Greenland, the Falkland Islands, or South Georgia and the other sub-Antarctic islands. You would look for them in vain on the frigid shores of Tierra del Fuego. But elsewhere they flourish as the dominant insects in terrestrial habitats of all kinds, from deserts to dense forests, thence to the fringes of the terrestrial world in marshland, mangrove swamps, and beaches. I have studied the three principal Arctic species above the tree line on Mount Washington in New Hampshire, where they are everywhere abundant, nesting under rocks to collect solar heat, and hurrying through one cycle of larval growth before the plunging temperature in September shuts their colonies down. Still, I have searched in vain for any ants above the tree line in New Guinea's Sarawaget Mountains, an inhospitable cycad savanna where cold rain closes in each day to soak all who try to stay there, whether human or formicid.

The eusocial insects are almost unimaginably older than human beings. Ants, along with their wood-eating equivalents the termites, originated near the middle of the Age of Reptiles, more than 120 million years ago. The first hominins, with organized societies and altruistic division of labor among collateral relatives and allies, appeared at best 3 million years ago.

To sense the difference, picture, if you will, a very distant ancestor of the first primates that were destined to be ancestors to humans, a small mammal scurrying about in search of dinosaur eggs through an early Cretaceous forest. As it climbs onto a coniferous log, a hind foot breaks through the bark. The interior is already partly hollow, the heartwood having been reduced into crumbling fragments by fungi, beetles, and a colony of primitive *Zootermopsis* termites. The cavity also serves as the nest for a colony of wasp-like sphecomyrmine ants. In a frenzy, the worker ants swarm over the offending mammal's leg, stinging any crevice or soft surface of the skin they can find. The animal, our ancestor, jumps off the log, shaking its leg and brushing off the attackers with a clawed foot. Had the cavity been occupied by a solitary wasp the size of a sphecomyrmine ant, the animal would scarcely have noticed it.

Now come forward a hundred million years to the present time. You, a descendant of the assaulted mammal, step onto a small pine log, the decaying trunk of a conifer descended from the one in the Cretaceous woodland. Descendants of the Cretaceous ter-

FIGURE 12-4. *A battle between ant colonies. Scouts from the nest (upper right),* Pheidole dentata, *colored black, have discovered invading fire ant workers,* Solenopsis invicta, *red in color, and engaged them. The most effective* Pheidole dentata *warriors are the large-headed soldiers, who use their powerful mandibles to dismantle the invaders. (Illustration © Margaret Nelson.)*

mite colony scuttle into a dark recess, a part of the cavity they occupy, just like their closely similar Mesozoic ancestors. The descendants of the ancient ant colony swarm out from another part of the same cavity to sting and repel you, also like their Mesozoic forebears. Together we are representatives of the two great hegemons of the terrestrial world. The difference is that the termites and ants had it all to themselves for a hundred million years, undisturbed until we ourselves finally inched up to the eusocial level.

The earliest ants arose from winged, solitary wasps. The workers of the first colonies evolved into creatures specialized for crawling on and under the ground and litter surface, and up from there onto living vegetation. At that point the workers flew no more. The virgin queens continued to fly, but each one only briefly, as they soared into the air and leaked sex pheromones to attract and mate with a winged male. Then they landed to start a new colony, never to fly again. Through further evolution the Mesozoic ants went on to build little civilizations by instinct, spreading their domains everywhere through the rotting vegetation on the surface and deep down into the soil beneath.

They evolved in complexity while proliferating new species dur-

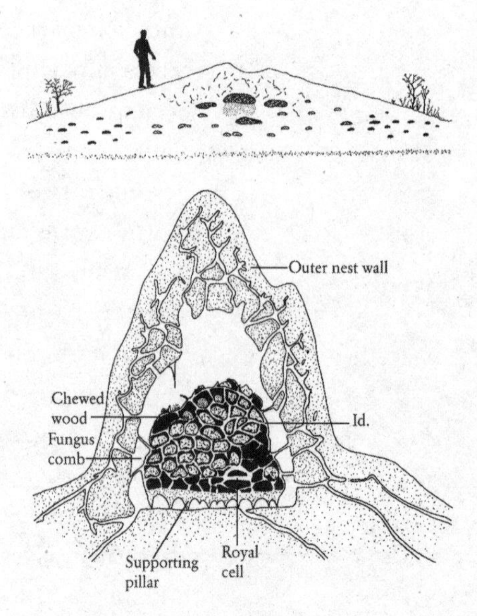

FIGURE 12-5. *Nests of the colony of the mound-building termites of the African genus* Macrotermes, *in cross section. The nest dissected in the upper panel was thirty meters in diameter. The nest dissected in the lower panel shows the architecture that creates air-conditioning. Air in the core is heated by the metabolism of the termites, causing it to rise and pass out of the upper mound exits, while fresh air is pulled in from subterranean channels located around the nest edges. The constant flow keeps the temperature, along with the oxygen and carbon dioxide levels, almost constant for up to the million termites living in the nest. (Modified from Edward O. Wilson,* The Insect Societies *[Cambridge, MA: Harvard University Press, 1971]. Based on research by Martin Lüscher.)*

ing tens of millions of years. Many became predators—the premier hunters of insects, spiders, sow bugs, and other ground-dwelling invertebrates—whose descendants still live with us today. Ants also took the role of primary undertakers, scavenging the remains of small animals killed by disease and accident. Of equally great importance to the whole terrestrial ecosystems, they became the preeminent turners of the soil, surpassing even the work of earthworms.

I have (very) crudely estimated the number of ants living today

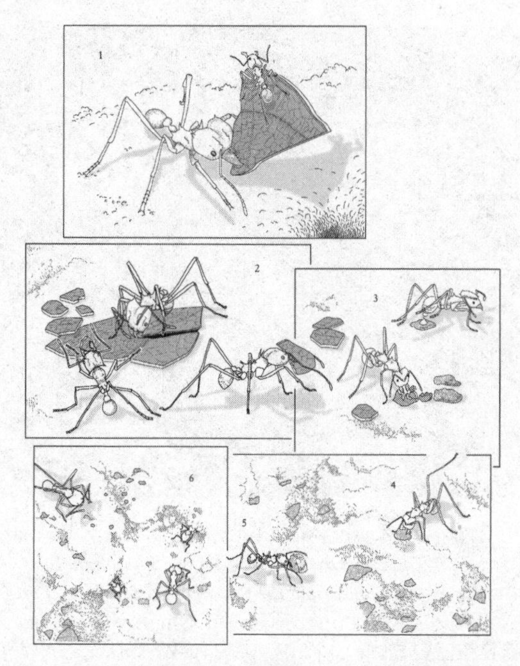

FIGURE 12-6. *The assembly line of the leafcutter ants, dominant insects of the American tropics, is the most complex social behavior of any known animal. (1) Large media workers find fresh vegetation, cut pieces off, and carry them to the nest; they are accompanied by tiny minimas that protect them from parasitic flies. (2) Inside the nest, smaller workers cut the pieces into 1-mm-wide fragments. (3) Still smaller medias chew the fragments into pulp. (4, 5) Minimas variously add pulp to the garden or tend the fungus growing there. (From Bert Hölldobler and Edward O. Wilson,* The Leafcutter Ants: Civilization by Instinct *[New York: W. W. Norton, 2011].)*

to be, at the nearest power of ten, 10^{16}, ten thousand trillion. If each ant on average weighs one-millionth as each human on average, then, because there are a million times more ants than humans (at 10^{10}), all the ants living on Earth weigh roughly as much as all the humans. This figure is not so impressive as it may sound. Consider: if every living person could be collected and log-stacked, we would make a cube less than one mile on each side. So if all the ants could

FIGURE 12-7. *Workers of the Australia weaver ant (*Oecophylla smaragdina*) build nests in the treetops by pulling leaves together to form chambers, then binding them in place with silk threads coaxed from the grublike larvae. (From Bert Hölldobler and Edward O. Wilson,* The Superorganism: The Beauty, Elegance, and Strangeness of Insect Societies *[New York: W. W. Norton, 2009]. Photo by Bert Hölldobler.)*

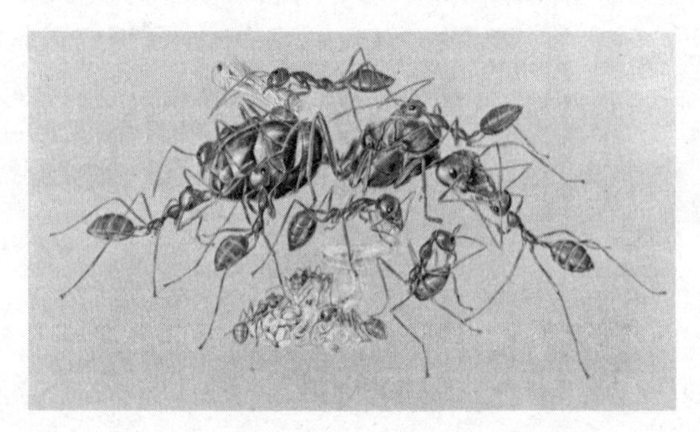

FIGURE 12-8. *Castes in a colony of African weaver ants (*Oecophylla longinoda*) include the queen, surrounded by major workers, who feed and groom her, and minor workers, who care for the grublike larvae, eggs, and pupae. Other major workers build aerial nests with silk threads contributed by the larvae. (From George F. Oster and Edward O. Wilson,* Caste and Ecology in the Social Insects *[Princeton, NJ: Princeton University Press, 1978]. Painting by Turid Hölldobler.)*

be similarly collected and log-stacked, they would make a cube of similar size. Both could be easily hidden in a small section of the Grand Canyon. Judged by protoplasm alone, they might seem less than an imperial spectacle. But what a piece of work are these two conquerors of Earth, ours to observe and compare.

· 13 ·

Inventions That Advanced the Social Insects

I WILL NOW TELL the story, which I helped to unravel during the past half century of research, of how the social insects rose to dominance among the invertebrates of the terrestrial world. These miniature conquerors did not burst like alien invaders into the environment. They insinuated themselves into it with quiet little steps, each taking millions of years to accomplish. At first they were ordinary, even rare elements in the Mesozoic forests and grasslands. Then they hit upon innovations in behavior and physiology parallel to human technological inventions. With the aid of each of their innovations, they entered new niches. Their ability to control the environment improved, and their numbers grew. By the middle of the Eocene period, 50 million years ago, they had become the most abundant of all medium-sized to large invertebrates on the land.

When ants first appeared, during the Late Jurassic period or Early Cretaceous period, termites had already flourished for tens of millions of years, but in a wholly different part of the same ecosystems. They were descendants of cockroach-like insects whose own ancestry dates back another hundred million years into the Paleozoic era. (I will pause to answer an oft-asked question: how can we tell termites, also called "white ants," from real ants? Easy, they have no waist.) Termites mastered the technique of digesting dead wood and other vegetation by forming symbioses—close biological partnerships—with lignin-degrading protozoans and bacteria living in their guts. After a very long period of time, some of the evolutionarily most advanced species created veritable cities by producing their food, like the leaf-cutting ants, in gardens of fungi grown on mulch, and by air-conditioning their nests. They divided labor among complicated arrays of physical castes.

In a sense, ants were to end up the more dominant of the two evolving lines, and mistresses over the twin insect empires, because many of their species became specialized to feed on termites, while no termite species ever learned to feed on ants. However, despite the greatness of their destiny, ants did not rush into immediate prominence upon their origin. For more than thirty million years, during the remainder of the Mesozoic era, they remained an ordinary presence surrounded by an immense variety of solitary insects. Other entomologists and I have searched through thousands of pieces of Mesozoic fossil resin (called amber) in search of these earliest ants. We have found them in the fossil beds of the right age in New Jersey, Alberta, Siberia, and Burma. We have come up with fewer than a thousand individuals, composing only a small minority among the other insects preserved in the same way. The specimens are spread over an age span of millions of years.

Fossils of ants this old were at first entirely unknown to scientists. For us the Mesozoic era, when the early history of these insects must have unfolded, was a complete blank. Then, in 1967, I received a piece of fossil metasequoia amber that two amateur collectors had picked up in a New Jersey stratum of Late Cretaceous age, about 90 million years old. Present together were two beautifully preserved worker ants in the transparent amber. They were almost twice as old as the most ancient ant fossil previously known. As I held the piece in my hand, I knew I was the first to look back into the deep history of one of Earth's two most successful insect groups. It was among the most exciting moments of my life (and I can understand if the reader does not appreciate my reaction to a fossil insect). In fact, I was so excited that I fumbled and dropped the piece. It fell to the floor and broke into two fragments. I froze and stared down in horror, as though I had just bumped into and shattered a priceless Ming

Dynasty vase. However, fortune continued to favor me that day. There remained one undamaged ant in each fragment, and each could be polished separately. As I studied these treasures closely, I found that their anatomy had traits intermediate between modern ants and wasps—one line of which must have been the ant ancestor. The hybrid nature was remarkably close to what a fellow researcher, William L. Brown, and I had earlier predicted. We gave the new species the name *Sphecomyrma*, meaning "wasp ant." Because of the eminence of ants in the world today (after all, the environment depends upon them), *Sphecomyrma* ranked in scientific importance with *Archaeopteryx*, the first such fossil intermediate between birds and their ancestral dinosaurs, and *Australopithecus*, the first "missing link" discovered between modern humans and the ancestral apes. The hunt was now on for additional Mesozoic ant fossils, to fill out a more complete history of these social insects.

As a subsequent intense search yielded more specimens, we also learned of changes occurring in the external environment that had made possible the rise of ants to eventual full dominance. Between 110 and 90 million years ago, still well back in Mesozoic times, the forests in which the ants lived began a profound transformation that made such an advance possible. Until that time, the trees and shrubs consisted primarily of gymnosperms, in particular the palmlike cycads, the ginkgos (today represented by a single species preserved as an ornamental), and, above all, the conifers, including pines, fir, spruce, redwood, and other "cone-bearers" (hence the name conifers) that still occur in forests scattered around the world. At the time the ants and termites entered the scene, the plant-eating dinosaurs were browsing on gymnosperms. Termites consumed the dead vegetation left over. Ants most likely excavated their nests in gymnosperm logs, in ground litter, and in the humus of the soil beneath. They searched the ground for food and climbed ferns and the canopies of the trees for food. Entomologists today are able to study a good number of specimens that were trapped in resin flows mostly of metasequoia trees, among the most abundant conifers of the Mesozoic era. Some of the fossils are beautifully preserved in this material, providing anatomical details that allow reconstruction of the early stages of ant evolution.

With the aid of the remains of many other kinds of animals and plants, I and other researchers have been able to reconstruct what happened next. Around 130 million years before the present time and peaking by 100 million years ago, one of the most radical and important changes in the history of life occurred. The gymnosperms were largely replaced by angiosperms, "flowering plants," which largely dominate the land environment today. Sequoias and their relatives gave way to the ancestors of magnolias, beech, and maple, and other familiar trees, while cycads and ferns yielded their dominance to grasses and the herbaceous angiosperms and shrubs of the ground flora.

Two evolutionary innovations during this time made the angiosperm revolution possible. First, endosperm in the seeds (the part we eat) made possible not only survival through unfavorable times but also long-distance dispersal. Second, the flowers and their attractive colors and scents allowed the evolution of an army of bees, wasps, flower flies, moths, butterflies, birds, bats, and other specialized creatures that transport pollen from the flower of one plant to the flower of other plants of the same species. Thus accoutered, the flowering plants spread around the world with relative swiftness (by geological standards). As their total range and abundance grew over millions of years, they filled the niches available to them while creating new ones with the bulk and complexity of their vegetation. More than a quarter million flowering plant species now exist on Earth, composing more than three hundred taxonomic families, including the very familiar Roseaceae (roses and relatives), Fagaceae (beeches), and Asteraceae (sunflowers and relatives). They are the tangled bank at the roadside, the meadows, the orchards, the croplands, and—by far the most diverse of all ecosystems—the tropical forests.

The ants were lifted on the tide of the flowering-plant evolution. The reason for the coevolution is, I am convinced, that the

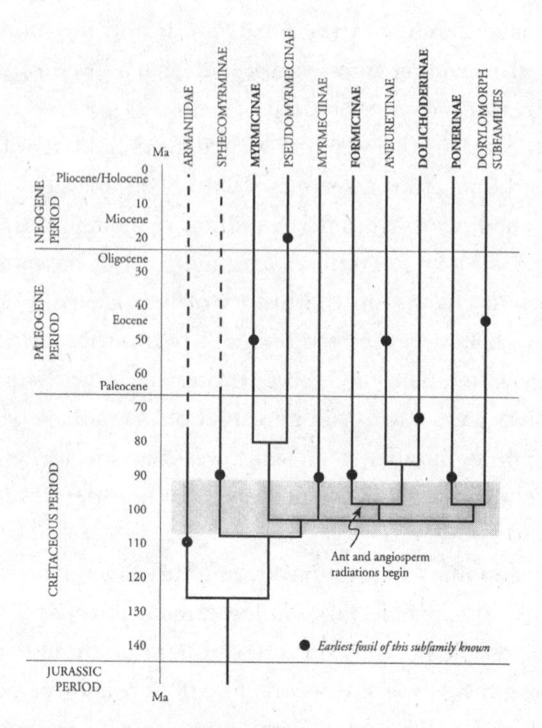

FIGURE 13-1. *In the Cretaceous period of the Age of Reptiles, the rise and diversification of ants still present today coincided with the domination of the Earth's flora by flowering plants (angiosperms). (From Edward O. Wilson and Bert Hölldobler, "The rise of the ants: A phylogenetic and ecological explanation,"* Proceedings of the National Academy of Sciences, U.S.A. *102[21]: 7411–7414 [2005].)*

angiosperm forests were richer in substance and more complicated in architecture, hence favorable to more kinds of small animals living in them. The undergrowth and fallen vegetation litter of the old gymnosperm forests in which ants had originated had been relatively simple in structure. As a result, fewer niches for insects and other small animals were available, and the variety of insects, spiders, centipedes, and other arthropods inhabiting the forests was proportionately smaller. The same relative paucity persists in the gymnosperm forests that have survived to the present time. The layers of litter and the soil beneath flowering plants of the new forests contained a far more complex environment for arthropods, including the ants that preyed upon them. The litter in which ant colonies of many species built their nests was more diverse in the kinds of decaying twigs, tree branches, clusters of leaves, and seed husks in which chambers and galleries could be excavated. In the angiosperm litter also was a greater range of temperature and humidity regimes encountered passing from top to bottom. For these reasons, a wider variation of arthropods were also available for food. The overall result was a global adaptive radiation of ants, with more and more species around the world able to specialize on both the nest site and the food they exploited. Species of ants multiplied, as more and more niches opened for them to occupy. By the end of the Mesozoic period, 65 million years ago, most of the two dozen taxonomic subfamilies of ants living today had come into existence.

Even with much of its diversity in place, however, the sprawling ant fauna did not immediately achieve the dominance in numbers of organisms and colonies it currently enjoys. The oldest fossils entomologists have turned up, preserved in both amber and rock fossils, are only moderately abundant in comparison with those of other insects. Possibly toward the end of the Mesozoic era ("Age of Reptiles") and certainly no later than the first 15 million years of the following Cenozoic era ("Age of Mammals"), the ants made two more evolutionary advances that today add to the basis of their world domination.

The first innovation was the strange partnership many of the species formed with insects that live on the sap of plants. Aphids, scale insects, mealybugs, and other members of the insect order Homoptera feed by piercing plants with their beaks and drawing up sap and other liquid materials. Each individual has to ingest a large amount of this substance in order to obtain enough nutrients to grow and reproduce. The constraint in their method of feeding requires that they also pass a large amount of excrement and excess liquid. The droplets are oozed or squirted out and allowed to fall to

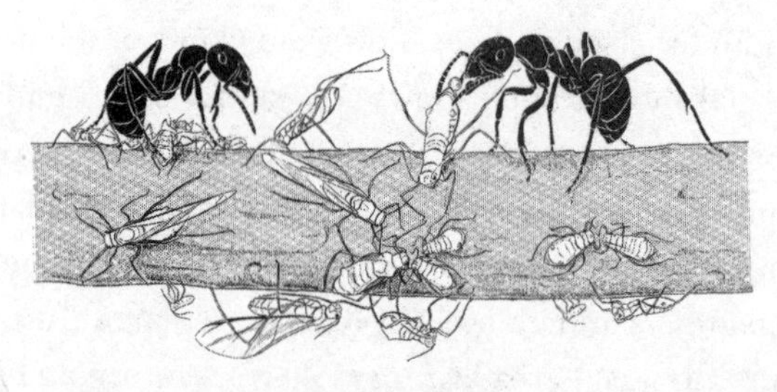

FIGURE 13-2. *A critical step in the rise of dominance of the ants is the partnerships they formed with sap-sucking insects, taking nutritious liquid excrement in exchange for protection against predators and parasites. This drawing is of the European ant* Formica polyctena *and its symbiotic aphid partner* Lachnus roboris. *(From Edward O. Wilson,* The Insect Societies *[Cambridge, MA: Harvard University Press, 1971]. Drawing by Turid Hölldobler.)*

the ground or surrounding vegetation, preventing the sticky material from piling up around the insects. Such "honeydew" is manna to most kinds of ants. For many species it is also a primary source of food.

The coming of the ants provided an equal advantage to its partners, and the symbiosis has endured to the present day. When their beaks pierce the plant epidermis, the aphids and other sapsuckers are literally anchored to their food. Their soft bodies provide tidy morsels for a host of predators and parasites that swarm through the foliage. Wasps, beetles, lacewings, flies, spiders, and others can wipe out the entire population on a plant in short order. The sapsuckers need constant protection, and an alliance with excrement-hungry ants is an excellent way to obtain it. Ants of many kinds treat any persistent rich food source as part of their territory, even if it is located far from their nests. They drive away any enemies from the herds of sapsuckers they claim as their own.

During their evolution, over millions of years, ants went further: they turned cooperative aphids and other sapsuckers into the equivalent of dairy cows. Or, put with equal accuracy, the sapsuckers turned ants into the equivalent of dairy farmers. For their part, the symbiotic sapsuckers stopped spritzing their excrement off the plant on which they rested, and simply held it in until an ant came along and touched them lightly with her antennae, whereupon the sapsucker extruded a generous drop and held it in place for the ant to drink. During their evolution, both partners of the symbiosis prospered. Others were not so fortunate. The plants lost a great deal of their plant blood, so to speak, and predators hunting for sapsuckers often went hungry. But all survived; such is an example of what is known as the balance of nature.

One day, on a hike through a New Guinea rainforest, I came upon a cluster of giant scale insects feeding on an understory shrub. Their bodies, encased in hard chitinous covers like turtle shells, were nearly ten millimeters across. Ants were in close attendance, scurrying about the herd, collecting droplets of honeydew. It occurred to me that these scale insects were big enough (or, looked at from a different perspective, I was small enough) for me to play the role of an ant. At the same time I was fortunately too big for the guardian ants to drive me away, although they tried. I plucked a hair from my head and touched its tip to the back of one of the scale insects—gently, as an ant might apply the tips of one of its own antennae. As I hoped, out popped a generous droplet of excrement. I picked it up with a pair of fine optician's forceps I carried, and tasted it. I found it mildly sweet. I also knew I was getting a small measure of amino acids that would have been good for my nutrition had I been an ant. To the scale insect, of course, I *was* an ant.

The ant-sapsucker partnership has been taken to extreme lengths during the geologically long association between the two kinds of insects. Many contemporary ant species manage their populations of six-legged cattle as all-purpose herds, eating some of them during periods of protein shortage. A few go so far as to carry them from worn-out pastures of vegetation to new, fresher ones. One species in Malaysia has even become a migratory herder, periodically moving its entire colony with its captive sapsuckers from place to place to obtain consistently high yields of honeydew.

Symbioses between ants and homopteran sapsuckers, as well as honeydew-secreting caterpillars of the butterfly family Lycaenidae ("blues"), are far from trivial curiosities. They occur in abundance around the world and are among the major links in the food chains that bind together many terrestrial ecosystems. For humans, they are important agricultural pests. For their part, the symbioses permitted the ants to occupy an entirely new dimension of the land environment. They had previously traveled up into the evergreen reaches of the tropical forests and returned to nest on or close to the ground. Now they could live all the time high above the ground. In many tropical regions, ants came to be the most abundant insects of the tree canopies.

For a long time, biologists were puzzled by the arboreal domination achieved by ants. How could such preeminently carnivorous creatures maintain such large populations? Their presence in great numbers at the top of the food chain seemed to violate a basic principle of ecology. Each gram of carnivore is supposed to consume many grams of herbivores (very roughly, ten times as much substance), as for example humans eating beef. The herbivores in turn feed on much larger masses of vegetation, as cattle upon grass.

When, finally, young and adventurous biologists climbed into the tropical canopies to observe the ant communities directly, they made an astonishing discovery. The ants are only part-time carnivores. To a large extent they are also herbivores. More precisely, they are *indirect* herbivores. The arboreal ants still can't digest vegetation on their own, the way caterpillars and scale insects do. That would require a major reengineering of their digestive systems. However, they can live off the nutritious excrement of sapsucker homopterans abounding in the treetops. The ants carefully protect and control herds of sapsuckers that build up in and around their nests. Some of the symbionts are maintained in "ant gardens," globular masses of epiphytic plants cultivated by the ants, such as orchids, bromeliads, and gesneriads. The gardens are both the homes and the pastures of the symbionts.

I have studied these garden ants myself in the rainforests of the Amazon and New Guinea—on the lowest tree branches, I confess, where no climbing was necessary. I was startled by their aggressiveness. Whenever I disturbed a nest, defending workers swarmed out to bite, sting, and spray poisonous secretions on whatever part of me they could reach. Quite possibly the most ferocious ant in the world on or above the ground is *Camponotus femoratus*, a medium-sized relative of the large black carpenter ant of the Northern Hemisphere, and abundant in South American rainforests. The garden-building *femoratus* I encountered did not allow me even to touch the nest. When I approached downwind to within several feet, the inhabitants smelled me. The workers swarmed out by the hundreds to form a seething carpet on the nest and began to spray mists of formic acid in my direction. When I persisted, they dropped on nearby vegetation to get closer. Anyone who has climbed onto the branches of a *femoratus*-inhabited tree needs no further explanation of the ecological dominance of ants.

In fierceness the Amazonian *Camponotus femoratus* is rivaled in equatorial Africa and Asia by the weaver ants of the genus *Oecophylla*. The colonies build nests of leaves pulled together by living chains of workers and sown in place by sheets of silk obtained, thread by thread, from the grub-like larvae of the colony. A mature colony constructs hundreds of these silken pavilions through the canopies of one to several trees. Any intruder into a weaver ant territory is met with bites and formic acid sprays from swarms of fearless defenders. When workers escaped from plastic cages in which I kept a colony at Harvard University, some would walk onto my desktop and threaten me with open mandibles, their abdominal tips lifted ready to spray me with formic acid. Their ferocity in the field is legendary. In the Solomon Islands during World War II, marine snipers climbing into trees were said to fear weaver ants as much as they did the Japanese. Hyperbole of course, but a tribute to the insects that rule Earth with us.

Over the years, I have come to recognize a principle relevant to our understanding of the evolutionary origin of the ants and other social insects: *the more elaborate and expensive the nest is in energy and time, the greater the fierceness of the ants that defend it.* This is a concept I will later connect to the origin of eusociality itself.

In roughly the same period of geological time that many kinds of ants were perfecting their partnership with honeydew-producing insects in the treetops, others were expanding their habitats and diets in an entirely different direction. To their basic menu of prey and carrion, they added seeds. The innovation permitted an increase in the number of species and density of colonies in the forest strongholds of the original ant faunas. It also allowed many kinds of ants to expand into arid grasslands and deserts.

Today many of the ant species that feed on seeds also build granaries in which to store them. The phenomenon occurs to a limited extent in forested areas, but was not perceived there or anywhere else until well into the nineteenth century, when naturalists began to study ants in the drier regions of the Levant, India, and western North America. Digging into the earthen nests of what came to be called "harvester ants," they found chambers packed with seeds of nearby herbaceous plants. Only then did the wisdom of Solomon make sense: "Go to the ant, O sluggard, observe her ways and be wise, which, having no chief, overseer or ruler, prepares the food in the summer and gathers her provision in the harvest."

One day, on a visit to Jerusalem's Temple Mount, I sat down close to a nest of harvester ants of the genus *Messor*, one of the dominant species of ants in the region. I watched as workers carried seeds down an entry hole on the way to the subterranean granaries. I entertained the conceit that this was likely the same species Solomon knew, and perhaps close to the same spot where he had seen them.

Three millennia later, and far from the land of Judaea, scientists have begun to turn to the ants and other social insects for a new kind of wisdom. Although these small creatures are radically different from us in many ways, their origins and history shed light upon our own.

IV

The Forces of Social Evolution

· 14 ·

The Scientific Dilemma of Rarity

EUSOCIALITY, THE CONDITION of multiple generations organized into groups by means of an altruistic division of labor, was one of the major innovations in the history of life. It created superorganisms, the next level of biological complexity above that of organisms. It is comparable in impact to the conquest of land by aquatic air-breathing animals. It is equivalent in importance to the invention of powered flight by insects and vertebrates.

But the achievement has presented a puzzle not yet solved in evolutionary biology: the rarity of its occurrence. For if one lucky population of wasps could give rise to the ants, and another lucky population of cockroach-like wood eaters turn into termites, and then the two of them dominate the land invertebrates, why hasn't the origin of eusociality been more common in the history of life? Why did it take so long in the history of life to occur?

The opportunities seem to have been superabundant. Before ants, termites, and social bees and wasps appeared on Earth, there were two massive and prolonged episodes of evolution by insects. The first began about 400 million years ago, during the Devonian period. It ended 150 million years later, at the close of the Permian period, when the greatest extinction of all time wiped out most species of plants and animals on Earth. Thus ended the Paleozoic—popularly known as the Age of Amphibians. It was succeeded by the Mesozoic era, the Age of Reptiles, both on the land and in the sea.

The Paleozoic era was the time of the coal forests, with tree ferns

FIGURE 14-1. *From the Middle through the Late Paleozoic era, about 400 million to 250 million years ago, insects of diverse kinds flourished on Earth. Their variety is illustrated by the array that could be found on a single tree fern, including beetles, cockroaches, and species of other extinct groups. None are known to have been social. (From Conrad C. Labandeira, "Plant-insect associations from the fossil record,"* Geotimes *43[9]: 18–24 [1998]. Drawing by Mary Parrish.)*

and towering scale trees. These forests and other terrestrial habitats scattered around them swarmed with insects, whose species rivaled in diversity those existing today. Present in abundance were ancient mayflies, dragonflies, beetles, and cockroaches. These familiar forms mingled with now extinct insects known only to experts who study their fossils—paleodictyopterans, protelytropterans, megasecopterans, diaphanopterodeans, and others given similarly unpronounceable names.

Pressed into fine-grained rock, many of the fossils are in remark-

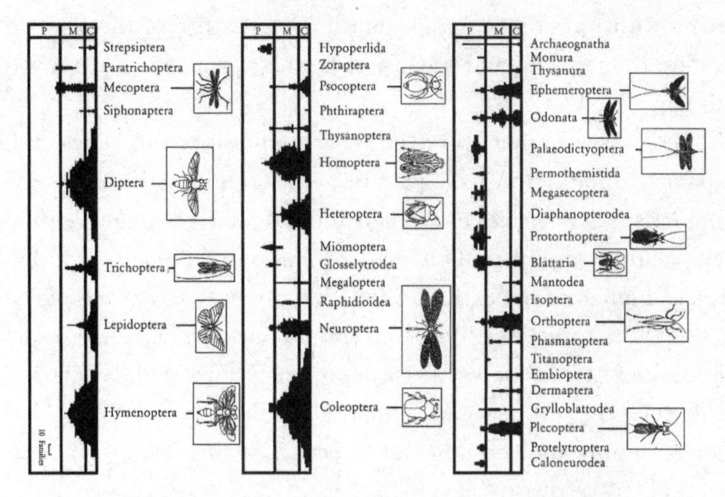

FIGURE 14-2. *Out of the vastness of insect diversity spanning 400 million years across three eras (Paleozoic, P; Mesozoic, M; Cenozoic, C), the origin of eusocial insects was very rare, and did not appear at all, so far as known, until the Early Mesozoic. The breadth of the diagrams represents the number of families in each insect order through time. (From Conrad C. Labandeira and John Sepkoski Jr., "Insect diversity in the fossil record,"* Science *261: 310–315 [1993]. Illustration prepared by Finnegan Marsh.)*

ably good condition, sufficiently so for us to compare most of their external anatomical details with those of modern insects. Researchers, using specimens collected from around the world, have been able to reconstruct the life cycles of some of the species, and even to deduce their diet. To this day, however, not a trace has ever been found of any eusocial insect.

There followed the great extinction that ended the Permian period and began the Triassic period, and with it the start of the Mesozoic era. Ninety percent of Earth's species were wiped out. Whatever caused this most catastrophic spasm of all time—most experts believe it would have been a mountain-sized meteorite, while others prefer internal events in plate tectonics or the chemistry of Earth itself—the episode came close to destroying plants and animals altogether. It did eliminate the aforementioned taxonomic orders with unfamiliar names, but it spared a few of the kinds of beetles, dragonflies, and other, less familiar groups that survive to this day.

The insects that survived the end-of-Permian extinction expanded rapidly (in geological terms) to refill Earth's land environments. Their species multiplied and radiated into many new lifeways. Within several million years, evolution of the survivors had replaced much of the extinguished diversity with new arrays of species, and the insect world became vibrant once again. Nevertheless, for another 50 million years, through much of the Triassic period, while the great evolutionary radiation of dinosaurs also unfolded, there still appeared no eusocial insects, at least none of which we can find any record.

Finally, in the latest part of the Jurassic period, some 175 million years ago, the first termites, primitively cockroach-like in anatomy, appeared, followed about 25 million years later by ants. Even then, and continuing to the present time, the origin of other eusocial insects, or eusocial animals of any kind, has been rare. Today there are approximately 2,600 recognized taxonomic families of insects and other arthropods, such as the common fruit flies of the family Drosophilidae, orb-weaving spiders of the family Argiopidae, and land crabs of the family Grapsidae. Only 15 of the 2,600 families are known to contain eusocial species. Six of the families are termites, all of which appear to have been descended from a single eusocial ancestor. Eusociality arose in ants once, three times independently in wasps, and at least four times—probably more, but it is hard to tell—in bees. Among the living eusocial sweat bees of the family Halictidae in particular, many lines are close to the very beginning of eusocial organization, with small colonies, barely differentiated queens, and a tendency to switch back and forth in evolution between the solitary and early eusocial states. These are the little bees, only a fraction the size of honeybees and bumblebees, that abound on asters and other kinds of flowers during the summer.

They are notably colorful: some are metallic blue or green, others banded black and white.

A single case of eusociality is known in ambrosia beetles, and others have been discovered in aphids and thrips. Amazingly, eusocial behavior has originated three times in shrimps of the genus *Synalpheus* of the family Alphaeidae, which build nests in marine sponges. Such rare or relatively unstable originations could easily have gone undetected in the fossil record. Also, the multiplicity of eusocial origins in the *Synalpheus* shrimps has been discovered only recently. A parallel caution has been raised by Geerat J. Vermeij from an analysis of twenty-three purportedly unique innovations in the mostly nonsocial aspects of life. Even with this uncertainty acknowledged, however, it is unlikely that many advanced and abundant eusocial insects, with their distinct worker castes, have gone entirely unnoticed.

Still rarer than in the invertebrates has been the appearance of eusociality in the vertebrates. It has occurred twice in the subterranean naked mole rats of Africa. It has occurred once in the line leading to modern humans, and in comparison with the invertebrate origins, only very recently in geological times—as recently as 3 million years ago. It is approached in helper-at-the-nest birds, in which the young remain with the parents for a time, but then either inherit the nest or leave to build one on their own. Eusociality is closely approached by African wild dogs, when an alpha female stays at the den to breed while the pack hunts for prey.

There were plenty of opportunities during the past 250 million years for such a momentous event as eusociality to occur in large animals. During Mesozoic times many evolving lines of dinosaurs attained at least some of the necessary prerequisites: human-sized, fast-moving carnivores, pack hunters, bipedal gait, and free hands. None took the final step to reach even primitive eusociality. For the next 60 million years, almost the entire duration of the Cenozoic era, the same opportunity lay before the proliferating species of large

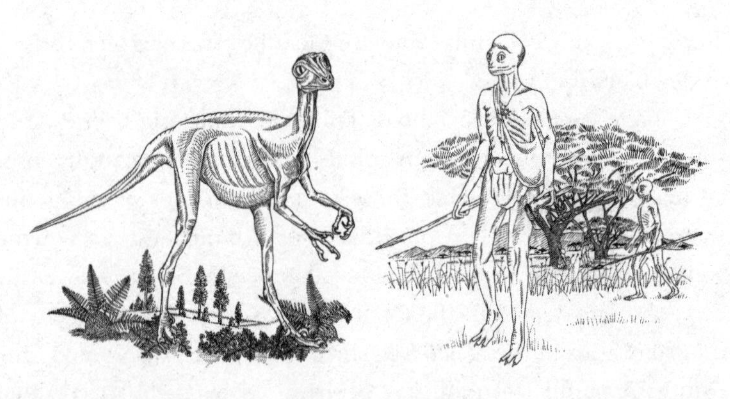

FIGURE 14-3. *What might have happened. On the left is a reconstruction of the bipedal dinosaur* Stenorhynchosaurus, *which lived near the end of the Mesozoic era and had some of the traits thought to make the origin of advanced intelligence possible. On the right is the "dinosauroid" as conceived by the paleontologist Dale Russell. This imaginary creature might have evolved from* Stenorhynchosaurus *a hundred million years before man—but did not. Based on an original reconstruction of* Stenorhynchosaurus *by Dale Russell. (From Charles Lumsden and Edward O. Wilson,* Promethean Fire: Reflections on the Origin of Mind *[Cambridge, MA: Harvard University Press, 1982].)*

mammals. Not only that, but the average life span of a mammal species and its daughter species averaged a comparatively short half million years, speeding the turnover in novel adaptations. Yet of all the nonprimate mammals in the world save the mole rats, and of all the primate species that lived across the tropical and subtropical regions for millions of years, only one, an offshoot of the African great apes, an antecedent of *Homo sapiens*, crossed the threshold into eusociality.

· 15 ·

Insect Altruism and Eusociality Explained

HUMANITY ORIGINATED AS a biological species in a biological world, in this strict sense no more and no less than did the social insects. What genetic evolutionary forces pushed our ancestors to the eusociality threshold, then across it? Only recently have biologists begun to solve this puzzle. Vital clues may be found in the histories of animal species, and especially the social invertebrates, that long before had blazed the same trail. The key, researchers discovered, was not to rely on any logical assortment of premises of what might have happened during the origin of the eusocial insects and other invertebrates, not to depend on mathematically constructed theories of what could have happened, but to piece together from field and laboratory observations what actually *did* happen. Cautiously, one step at a time, we have begun to piece together this story out of empirical evidence. The basic principles of genetics and evolution adduced might then be used, tentatively in the best spirit of science, to address the human condition.

The beginnings of a solid reconstruction of the invertebrate story, especially that of insects, was made in the middle of the last century by several great entomologists, William M. Wheeler, Charles D. Michener, and Howard E. Evans. As a younger scientist, I knew Michener and Evans personally very well (Michener is still alive and active in 2012), and although Wheeler died in 1937, when I was still a little boy, I have studied his research so closely and heard so much about his life since that I feel as though I also personally knew him.

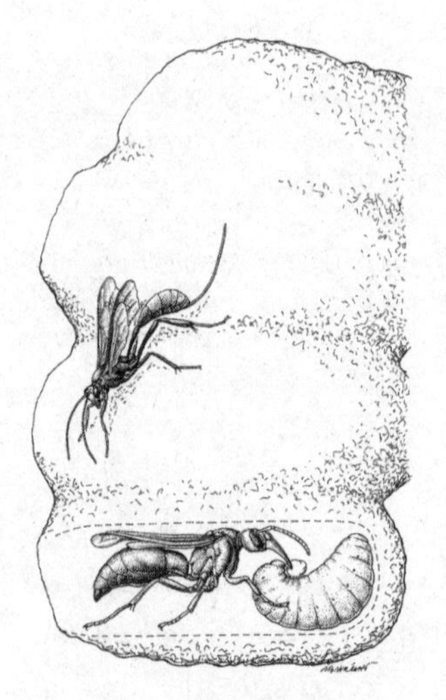

FIGURE 15-1. *Progressive provisioning in a solitary wasp. A cutaway view of a nest shows a female* Synagris cornuta *feeding her larva with a fragment of caterpillar. A parasitic ichneumonid wasp,* Osprynchotus violator, *lurks on the outside of the nest, waiting for the right moment to attack the larva. (David P. Cowan, "The solitary and presocial Vespidae," in Kenneth G. Ross and Robert W. Matthews, eds.,* The Social Biology of Wasps *[Ithaca, NY: Comstock Pub. Associates, 1991].)*

The three men were authentic naturalists of a kind much needed today on the frontiers of biology. Their scientific careers were devoted to learning everything there is to know about the group of organisms on which they specialized. Each became a world authority—Michener on bees, Evans on wasps, and Wheeler on ants. The center of their passion was the science of classification, but they also ventured beyond, to the ecology of their chosen subjects, to anatomy, to life cycles, to evolutionary relationships, to behavior. If you were fortunate enough to go into the field with one of the three, he could give you the scientific name of every bee (Michener), wasp (Evans), and ant (Wheeler) encountered, and he would relate with enthusiasm all that had been learned about the species up to that time. *Each had a feel for the organism*—and that is what mattered.

The mass of biological knowledge accumulated by many such scientific naturalists working in the field and laboratory has made it possible to develop a clear picture of how and why eusociality, the most advanced state of social behavior, came into existence. The sequence had two steps. First, in all of the animal species that have attained eusociality—all of them, without known exception—altruistic cooperation protects a persistent, defensible nest from enemies, whether predators, parasites, or competitors. Second, this step having been attained, the stage was set for the origin of eusociality, in which

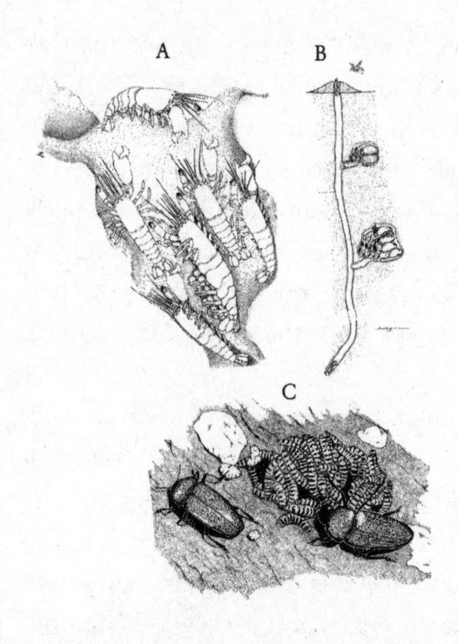

FIGURE 15-2. *Species on either side of the eusociality threshold. (A) Colony of a primitively eusocial* Synalpheus *snapping shrimp, occupying a cavity excavated in a sponge. The large queen (reproductive member) is supported by her family of workers, one of whom guards the nest entrance (from Duffy). (B) A colony of the primitively eusocial halictid bee* Lasioglossum duplex, *which has excavated a nest in the soil (from Sakagami and Hayashida). (C) Adult erotylid beetles of the genus* Pselaphacus *leading their larvae to fungal food (from Costa); this level of parental care is widespread among insects and other arthropods, but has never been known to give rise to eusociality. These three examples illustrate the principle that the origin of eusociality requires the preadaptation of a constructed and guarded nest site. (J. T. Costa,* The Other Insect Societies *[Cambridge, MA: Harvard University Press, 2006]; J. Emmett Duffy, "Ecology and evolution of eusociality in sponge-dwelling shrimp," in J. Emmett Duffy and Martin Thiel, eds.,* Evolutionary Ecology of Social and Sexual Systems: Crustaceans as Model Organisms *[New York: Oxford University Press, 2007]; S. F. Sakagami and K. Hayashida, "Biology of the primitively social bee,* Halictus duplex *Dalla Torre II: Nest structure and immature stages,"* Insectes Sociaux *7: 57–98 [1960].)*

members of groups belong to more than one generation and divide labor in a way that sacrifices at least some of their personal interests to that of the group.

To envision the process in a concrete manner, consider a solitary wasp who builds a nest where she raises her young. This is the step reached by birds and crocodilians. In the life cycle of the ordinary wasp species, the young leave the nest when they mature, and disperse to breed and build nests on their own, as do, for example, birds and crocodilians. If at least some of the next generation stay at the nest instead of dispersing, the resulting group has reached the eusociality threshold. That barrier is then easily crossed—albeit far from easily sustained thereafter. Bees of at least some solitary species (and communal bees that occupy a common burrow but build private cells) can be converted to the primitively eusocial state simply by placing two bees together in a space so small that only one nest or private cell can be built. The pair automatically form a pecking order of the kind observed in natural populations of primitively eusocial bees. The dominant female, the "queen," stays at the nest and reproduces and guards the nest, while the subordinate female, the "worker," forages for food.

In nature the same arrangement can be genetically programmed, with the mother insect surrounded by her offspring remaining at the nest, so the mother becomes queen and the offspring become workers. The only genetic change needed to attain the final step is the acquisition of an allele—one new form of a single gene—that silences the brain's program for dispersal and prevents the mother and her offspring from dispersing to create new nests.

As soon as such a cohesive group comes into existence, natural selection acting at the level of the group begins. This means that an individual in a group capable of reproduction does better, or worse, than an otherwise identical solitary individual in the same environment. What determines the outcome is the emergent traits due to the interactions of its members. These traits include cooperation in expanding, defending and enlarging the nest, obtaining food, and rearing the immature young—in other words, all the actions a solitary, reproducing insect would normally perform on her own.

When the allele prescribing the foregoing emergent traits of the group prevails over competing alleles that prescribe dispersal by individuals from the nest, natural selection on the rest of the genome is set free to create more complex forms of social organization. In the earliest stages of eusocial evolution, it nonetheless first acts upon the already existing predisposition to dominance and division of labor. Later, more of the remainder of the genome (that is, the whole genetic code) can participate at the level of the group, creating increasingly complex societies.

In the old, conventional image, that of kin selection and the "selfish gene," the group is an alliance of related individuals that cooperate with one another because they are related. Although potentially in conflict, they nonetheless accede altruistically to the needs of the colony. Workers are willing to surrender some or all of their personal reproductive potential this way because they are kin and share genes with them by common descent. Thus each favors its own "selfish" genes by promoting identical genes that also occur in its fellow group members. Even if it gives its life for the benefit of a mother or sister, such an insect will increase the frequency of genes it shares with the relatives. The genes increased will include those that produced the altruistic behavior. If other colony members behave in similar manner, the colony as a whole can defeat groups composed of exclusively selfish individuals.

The selfish-gene approach may seem to be entirely reasonable. In fact, most evolutionary biologists had accepted it as a virtual dogma—at least until 2010. In that year Martin Nowak, Corina Tarnita, and I demonstrated that inclusive-fitness theory, often called kin selection theory, is both mathematically and biologically incorrect. Among its basic flaws is that it treats the division of labor between the mother queen and her offspring as "cooperation," and their dispersal from the mother nest as "defection." But, as we pointed out, the fidelity to the group and the division of labor are not an evolutionary game. The workers are not players. When eusociality is firmly established, they are extensions of the queen's phenotype, in other words alternative expressions of her personal genes and those of the male with whom she mated. In effect, the workers are robots she has created in her image that allow her to generate more queens and males than would be possible if she were solitary.

If this perception is correct, and I believe it is in both logic and fit to the evidence, the origin and evolution of eusocial insects can be viewed as processes driven by individual-level natural selection. It is best tracked from queen to queen from one generation to the next, with the workers of each colony produced as phenotypic extensions of the mother queen. The queen and her offspring are often called superorganisms, but they may equally be called organisms. The worker of a wasp colony or ant colony that attacks you when you disturb its nest is a product of the mother queen's genome. The defending worker is part of the queen's phenotype, as teeth and fingers are part of your own phenotype.

There may immediately seem to be a flaw in this comparison. The eusocial worker, of course, has a father as well as a mother, and therefore partly a different genotype from that of the mother queen. Each colony comprises an array of genomes, while the cells of a conventional organism, being clones, compose only the one genome of the organism's zygote. Yet the process of natural selection and the single level of biological organization on which its operations occur are essentially the same. Each of us is an organism made up of well-integrated diploid cells. So is a eusocial colony. As your tissues proliferated, the molecular machinery of each cell was either turned on or silenced to create, say, a finger or a tooth. In the same way, the eusocial workers, developing into adults under the influence of pheromones from fellow colony members and other environmental cues, are directed to become one particular caste. It will perform one or a sequence of tasks out of a repertory of potential performances hardwired in the collective brains of the workers. For a period of time, rarely throughout its life, it is a soldier, a nest builder, a nurse, or an all-purpose laborer.

Of course, it is a fact that genetic diversity of traits among the workers of eusocial colonies not only exists but functions on behalf of the colony—as documented for disease resistance and climate control of the nest. Would this make the colony a group of individuals, each of whom (in the perspective of kin selection theory) seeks to maximize the fitness of its own genes? That such need not be the case becomes apparent if one views the queen's genome as consisting of parts relatively low in the variety of its alleles (different forms of each gene) whenever the traits they prescribe need to be inflexible, and yet in the same genome other parts are high in the variety of its alleles whenever those traits need to be flexible. Genetic inflexibility is a necessity of worker caste systems and the means by which they are organized and their personal labor distributed. In contrast, genetic flexibility in worker response is favored in disease resistance by the colony and in climate control inside the nest. The more genetic types that exist in a colony, the more likely that at least a few will survive if a disease sweeps through the nest. And the greater the breadth of sensitivity in detecting deviations from the desired temperature, humidity, and atmosphere, the closer these components of the nest environment can be held to their optimum for life of the colony.

There is no important genetic difference between the queen and her daughters in the potential caste they can become. Each fertilized egg, from the moment the queen and male genomes unite, can become either a queen or a worker. Its fate depends on the particularities of the environment experienced by each colony member during its development, including the season in which it is born, the food it eats, and the pheromones it detects. In this sense the workers are robots, produced by the mother queen as ambulatory parts of her phenotype.

In social hymenopteran colonies (ants, bees, wasps) that are "primitively" simple, in other words with few anatomical differences between the queen and her worker progeny, a state of conflict often results when workers try to reproduce on their own. The other workers typically thwart the usurpers, thus protecting the queen's primacy. They may just drive her away from the brood chamber whenever she tries to lay eggs. They may pile on the offender to punish her, perhaps severely enough to cripple or kill her. If she manages to sneak her eggs into the brood chamber, her co-workers recognize their different odor and remove and eat them. Many studies have shown that the degree of such conflict is correlated with the genetic difference between the would-be usurpers and the queen. Some of this phenomenon might be explained by a genetically based difference in odor, which then determines the degree of antagonism. Even so, the question remains whether such conflict is evidence against individual-level, queen-to-queen natural selection. That is not the case if the usurpers are viewed as parallels to cancer cells in the mammalian organism. The complex cellular apparatus of mammals, entailing T-cells, T-cell receptors, B-cell manufacture, and the major histocompatibility complex, serves the same functions—resisting infection and runaway cell growth—as does genetic variability among offspring of the queen.

Group selection occurs, in the sense that success or failure of the colony depends upon how well the collectivity of the queen and her robotic offspring does in competition with solitary individuals and other colonies. Group selection is a useful idea in identifying precisely the targets of selection when queens (and their colonies about them) are competing with other queens. But multilevel selection, in which colonial evolution is regarded as the interests of the individual worker pitted against the interests of its colony, may no longer be a useful concept on which to build models of genetic evolution in social insects.

Further, the very idea of altruism within an insect colony, while a nice metaphor, turns out to have little analytic value in science. If the object of interest is altruism in the sense of the sacrifice of personal reproduction, the goal of explaining it by multilevel-selection theory is likely to be illusory. The mother, her genes screened by individual selection, has the power to create workers to further her Darwinian fitness. Take away the power, and she fails.

Remarkably, Darwin hit upon the same basic concept in *The Origin of Species*, although in rudimentary form. He had thought long and hard about the problem of how sterile ant workers could evolve by natural selection. The difficulty, he worried, "at first appeared to me insuperable, and actually fatal to my whole theory." Then he solved the puzzle with the concept we call today phenotypic plasticity, with the mother queen and her progeny together as the target of selection by the external environment. The ant colony is a family, he suggested, and "selection may be applied to the family, as well as to the individual, and may thus gain the desired end. Thus, a well-flavoured vegetable is cooked, and the individual is destroyed; but the horticulturist sows seeds of the same stock, and confidently expects to get nearly the same variety. . . . Thus I believe it has been with social insects: a slight modification of structure, or instinct, correlated with the sterile condition of certain members . . . of the fertile males and females of the same community flourished, and transmitted to their fertile offspring a tendency to produce sterile members having the same modification."

The well-flavored vegetable is a nice metaphor. The superorganism is the queen, with her servant daughters busy about her. With modern biology we can now explain, I believe, how such a creature came into being.

· 16 ·

Insects Take the Giant Leap

I WILL NOW PRESENT a scientific argument simplified for a general readership, but also constructed in the style appropriate for a technical subject still in rapid development, with several topics in it still subject to challenge.

From Darwin to the present time, the study of eusocial origins and evolution has focused on the large assemblage of species belonging to the Hymenoptera, the insect taxonomic order that includes the ants, bees, and aculeate (stinging) wasps. More distantly related assemblages within the Hymenoptera are the parasitoid wasps and nonparasitic sawflies and horntails, creatures that swarm all around us in nature but are seldom noticed. By scanning the natural histories of thousands of species of these insects, entomologists have pieced together the finely graded steps in evolution that evidently led from solitary individuals to advanced, eusocial colonies. This knowledge when arrayed in logical steps leading to eusociality contains clues to the genetic changes and forces of natural selection by which each step in turn was achieved.

One solid principle drawn from this analysis of the hymenopterans, and other insects as well, is that all of the species that have attained eusociality, as I have stressed, live in fortified nest sites. A second principle, less well established but probably nonetheless universal, is that the protection is against enemies, namely predators, parasites, and competitors. A final principle is that, all other things being equal, even a little society does better than a solitary individual belonging to closely related species both in longevity and in extracting resources from the area around a fixed nest of any kind.

The resource exploited in early stages leading to eusociality in all known cases consists of a nest guarded by workers and within foraging range of a dependable food source. To take one well-studied stage, the females of a great many aculeate wasps, such as mud daubers and spider wasps, build nests and then provision them with paralyzed prey for the larvae to consume. Among the 50,000 to 60,000 aculeate species known from around the world, at least seven independently evolving lines have gone on to attain eusociality. In contrast, among the more than 70,000 known parasitic and other nonstinging hymenopteran species, whose females travel from prey to prey to lay their eggs, none is eusocial. Nor is any known among the hugely diversified 5,000 described species of sawflies and horntails. Such is even the case of the many sawfly species that form well-coordinated aggregations. They may seem to be on the cusp of eusociality; they may seem to be only one simple mutation away. But none has passed over to it; none has a queen and worker castes.

Outside the Hymenoptera, all of the thousands of known species of bark and ambrosia beetles, which compose the taxonomic families Scolytidae and Platypodidae, depend on dead wood for shelter and food. Many of these tiny insects also dig burrows and care for their young in them. A very few are able to cut and sustain burrows in living heartwood, allowing the coexistence of individuals across multiple generations. Among the latter only one, the Australian eucalyptus-boring beetle *Platypus incompertus*, is known to have developed eusociality. Because of the persistence of this species' habitat, tunnel systems are estimated to have survived, and presumably have housed the same families, generation after generation for up to thirty-seven years.

In a parallel manner, the handful of aphid and thrips species known to be eusocial all induce galls. The swollen tumor-like growths are found in a wide variety of plants. If you are ever curious about the meaning of galls, cut a fresh one open on living vegetation and inside you will usually find the insect that caused it. The aphid and thrips colonies occupy cavities within the galls, enjoying a rich food supply in a secure, defensible home of their own making. In contrast, the vast majority of other known species of aphids and the closely related adelgid species, roughly 4,000 in number, and thrips, about 5,000 strong, often form dense aggregations, but do not cultivate galls or divide labor.

In shallow marine waters of the American tropics, several species of the shrimp genus *Synalpheus,* out of roughly 10,000 known described decapod crustacean species in the world, have uniquely reached the eusocial level. *Synalpheus* shrimp are also highly unusual among decapods in excavating and defending nests in sponges.

A second trait that originates in solitary ancestors but predisposes species to evolve eusocial colonies has been documented in sweat bees of the taxonomic family Halictidae. When researchers experimentally forced together two solitary bees in the halictid genus *Ceratina* and *Lasioglossum,* the coerced insects proceeded in repeated such trials to divide labor variously in nest-building, foraging, and guarding. Furthermore, in at least two species of *Lasioglossum,* females engage in leading by one bee and following by the other bee. The same interaction routine characterizes primitively eusocial species.

This surprising anticipation of social behavior in solitary bees that has no apparent Darwinian rationale appears instead to be the result of a preexisting ground plan that guides the labor and life cycle in solitary species. In the ground plan, solitary individuals tend to move from one job to another after the first is completed. In eusocial species, this simple algorithm of labor is transferred to the avoidance of a job already completed or being filled at the time by a nestmate. The result is a more even spread of labor as needs in the colony open up.

Thus solitary but progressively provisioning bees are spring-loaded—that is, strongly predisposed, and provided as with a trigger—for a rapid evolutionary shift to eusociality, once natural selection favors the division of labor that characterizes eusociality.

At the next lower level of biological cause and effect, built into the way the nervous system itself works, we find a likely explanation of the spring-loading of early social behavior. The self-organization of two solitary bees forced together fits the "fixed-threshold" model of the origin of labor division in eusocial species. The fixed-threshold model posits that variation, sometimes genetic in origin among individuals and sometimes not, exists in the amount of stimulation needed to trigger work on particular tasks. When two or more individual ants or bees together encounter the same available task, those with the lowest amount of stimulation needed are the first to begin working. The activity inhibits their partners, who are then more likely to move on to whatever other tasks are available. Thus, once again, a simple change in the nervous system, which this time is due to a substitution of one allele with a flexible outcome in its effect, could be enough to carry a preadapted species across the threshold to eusociality.

For a solitary animal species, to be near the eusociality threshold means to be engaged in progressive provisioning of a defensible nest. The approach to the threshold is attained in a happenstance manner by conventional natural selection at the individual level. Whether a eusocial allele proves successful and spreads through the population is an accident: its fate depends on whether the particular environment around the nest is of a kind that favors eusocial groups over individuals.

When all the necessary conditions occur—namely the right pre-eusocial traits are in place, a eusocial allele also exists in the population, even if at very low levels, and, finally, environmental pressures exist that favor group activity—the solitary species will move across the threshold into eusociality. The surprising aspect of this evolutionary step is that the eusociality gene does not need to create new forms of behavior. As in the case of many random mutations generally, it need only silence a preexisting behavior, thus halting the dispersal of parents and grown offspring from the nest.

As a result of the cancellation, the family stays home. Looking at the matter the other way, the eusociality gene they share with the mother queen has turned them into robots, expressing one state of her own flexible phenotype. In this sense, I have argued, the primitive colony is a superorganism. It is essentially a kind of organism in which the working parts are not the usual cells but pre-subordinated organisms.

Eusociality and what we like to call altruism can be born of the flexible expression of a single allele (gene form) or ensemble of alleles, whenever parents were already building nests and feeding their young progressively. The only thing needed is group selection, acting on group traits that also favors families that stay at home. Then the advance to ecological dominance can begin. A new level of biological organization is attained. *One small step for a queen with her newly created worker caste, one giant leap for the insects.*

The shift to the eusocial level comes ultimately from the pressures put on the mother and her little colony from the external environment. What exactly are these environmental pressures? Field and laboratory research on this subject has scarcely begun, but a few suggestive examples have been worked out—providing a little part of the larger picture, a glimmering of what may be the true story. For example, females of the solitary nest-building wasp *Ammophila pubescens* provision their soil burrows with caterpillars, creating cells in the same burrow in succession, one on top of the other. Forced to open and close the nests inside each time, they lose many of their eggs to parasitic cuckoo flies that constantly patrol the area. It is entirely reasonable to suppose that if a second *Ammophila* female were available to serve as a guard, the loss of eggs would be considerably reduced. If the pair were further able to switch to progressive provisioning, in which the larvae hatching from the eggs could be raised on caterpillars brought to them as they grew up, and if the mother and adult offspring remained at the same nest, eusociality would be achieved.

Concrete examples of this adaptation and the transition it affords are provided by the primitively eusocial halictid sweat bees and polistine wasps. In one suggestive case recently worked out by researchers, two species of sweat bees that switched from collecting the pollen of many plant species to collecting pollen from only a few plant species, also reverted from a primitively eusocial life back to a solitary life. The explanation for this shift turns out to be self-evident. Specialization on a limited number of plant species is common among insects when it allows them to outcompete other plant-eating insects. Such a change in life history, which is presumably genetic in origin, also shrinks the length of the harvesting season and removes the possibility of overlapping generations—hence the formation of a eusocial colony and the advantage that might accrue from the presence of guard bees.

Evolution in the reverse direction is easily conceivable, and very likely has occurred. An adaptation to a broader array of food plants sets the stage for multiple generations, and thence for overlapping generations in the same nest. Similar evidence with respect to overlapping generations has been obtained for primitively eusocial wasps. In crossing the line to eusociality, a single allele that disposes daughters to stay can be fixed in the populations at large if the advantage of the little group over solitaires outweighs the advantage of each offspring leaving to try on its own. When this happens, the queen in effect switches from producing daughters that disperse to producing robotic helpers. The prescription is flexible: in the mating season some of the female offspring can be raised as virgin queens programmed to disperse and start new colonies.

The final step to eusociality, the addition of only one allele or a small set of alleles that silences the genes prescribing dispersal from the mother nest, is a distinct possibility in the real world. Throughout the great diversity of living ant species, for example, the coexistence of winged reproductive females and wingless worker females is a basic trait of colonial life. Judging from the flies (order Diptera) and butterflies (order Lepidoptera), both ancient groups, wing development is directed throughout the winged insects by an unchanged regulatory gene network. As much as 150 million years ago, the earliest ants (or their immediate ancestors) altered the regulatory network of wing development in such a way that some of the genes could be shut down under the influence of diet or some other environmental factor. Thus was produced a wingless worker caste.

An equally informative example of a small genetic change amplified downstream into a greater social change is the one affecting queen number and territorial behavior in the imported fire ant *Solenopsis invicta.* Colonies of the early U.S. population, which descended from colonies introduced by cargo out of southern South America by the mid-1930s, each contained one or a small number of functioning queens. The colonies also displayed odor-based territorial behavior, causing nests built by different colonies to spread out. Sometime during the 1970s, this strain of fire ants began to yield to another strain, whose colonies possess many queens and no longer defend territories. It turns out that the differences between the two strains are due to variation in a single major gene, *Gp-9.* The two *Gp-9* alleles have been sequenced, and their product appears to be a key molecular component engaged in the olfactory recognition of nestmates. The effect of the many-queen allele is evidently to reduce or knock out the ability to discriminate nestmates from members of other colonies, as well as to discriminate among potential egg-laying queens. As a result of the latter effect, colonies lose an important means of regulating queen number, with profound consequences for colony organization.

The exact nature of the genetic step to the earliest degree of eusociality remains unknown, unlike the cases of winglessness and colony odor, but it is immediately accessible to future genetic research. Biologists have suggested that the genetic base of the flexible worker-versus-queen difference in *Polistes* paper wasps is the same as the genetically based developmental physiology that regulates hibernation in solitary Hymenoptera. Such a shift in response to the environment may indeed be important. Oddly, the change need not be an allele or ensemble of alleles that appears by mutation and then spreads from low frequencies by group selection. Instead, the key allele may be previously fixed in the population by indi-

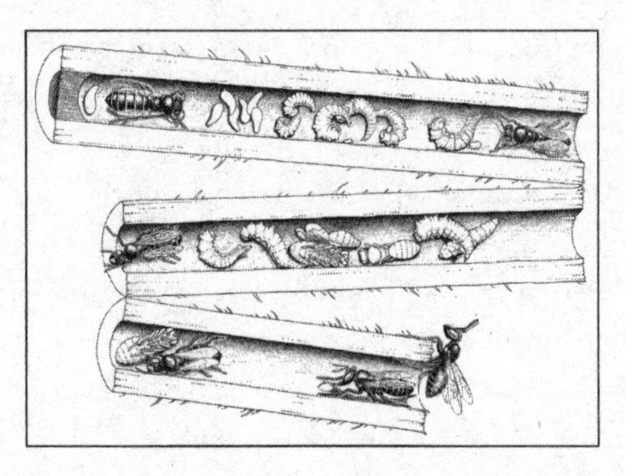

FIGURE 16-1. *A colony of a primitively eusocial Formosan bee (*Braunsapis sauteriella*) nesting in a hollow* Lantana *stem. The queen, with giant eggs, is to the left in the top segment. The workers feed the grublike larvae progressively with lumps of pollen, which are placed on the stem cavity walls. (From Edward O. Wilson,* The Insect Societies *[Cambridge, MA: Harvard University Press, 1971]. Drawing by Sarah Landry, based on an illustration by Kunio Iwata in Sakagami, 1960.)*

vidual direct selection rather than by group selection, with solitary behavior the norm in most environments and eusocial behavior in other, rare and extreme environments. With a shift in the available environment in space or time, eusocial behavior would become the norm. The potential of a species on the brink of eusociality to follow this path is shown by the Japanese stem-nesting xylocopine bee *Ceratina flavipes.* The vast majority of the females provision their nests with pollen and nectar as solitary foundresses, but in slightly more than 0.1 percent of the nests, two individuals cooperate. When this happens, the pair divides the labor: one lays the eggs and guards the nest entrance while the other forages.

Another example of genetic flexibility at the eusociality threshold is provided by the ground-nesting halictid sweat bee *Halictus sexcinctus.* The species is balanced on the knife edge of social evolution. In southern Greece, colonies of one hereditary strain are founded by cooperating females, and those of a second strain are founded by a single, territorial female whose offspring serve as workers.

Although some individual direct selection may play a role in the origin of eusociality, the force that targets the maintenance and elaboration of eusociality is by necessity environmentally based group selection, which acts upon the emergent traits of the group as a whole. An examination of the behavior of the most primitively eusocial ants, bees, and wasps shows that these traits initially include dominance behavior, as well as reproductive division of labor, plus, very likely, some form of alarm communication based on the release of pheromones. A species in the earliest stage of eusociality, to repeat for emphasis what I argued earlier, is a genetic chimera. On the one hand, the traits newly emerged in eusociality favor the group, while much of the rest of the genome, having been the target of individual direct selection over millions of years prior to the eusociality event, favors personal dispersal and reproduction. In order for the binding effects of group selection to outweigh the dissolutive effects of individual direct selection, the candidate insect species must have only a very short evolutionary distance to travel, such that no more than a small number of emergent traits are needed to form a eusocial colony. The reduction of that distance is achieved by a particular set of preadaptations, including the construction of a nest in which offspring are reared. The relative rarity of these preadaptations, when added to the high bar to eusociality set by countervailing individual direct selection, may be enough to explain the rarity of eusociality that exists throughout the history of the animal kingdom.

The only genetic change needed to cross the threshold to the eusocial grade is possession by the foundress of an allele that holds the foundress and her offspring to the nest. The preadaptations provide the flexibility in body form and behavior required for eusociality, as well as the key emergent traits arising from interactions of the group members. Group (colony-level) selection then immediately begins to act on both of these traits. The potential for an extreme elaboration of social organization is present, and it has in fact been achieved many times in the ants, bees, and termites.

In the earliest stage of eusociality, the offspring remaining in the nest would be expected to assume the worker role, in conformity with the preexisting behavioral ground rule inherited from the pre-eusocial ancestor. Subsequently, a morphological worker caste (distinguished from the larger, fertile queen caste) can emerge by a further genetic change in which the expression of genes for maternal care is rerouted to precede foraging, thus reversing the normal sequence in the adult developmental ground plan of the ancestor. The rerouting is programmed to retain part of the phenotypic plasticity of the alleles that prescribe the overall ground plan. This origin of an anatomically distinct worker caste appears to mark the "point of no return" in evolution, at which eusocial life becomes irreversible. If the colony royals could talk, they might then say, in pheromone language, "We will all stand together, on every one of our six legs, or we will fall together." There must be balance and cooperation. If too many queens, there will not be enough workers to maintain the colony. Too many workers, and food around the nest will fall short. Not enough soldiers, and predators will overwhelm the nest. Not enough foragers venturing outside the nest, and the colony will starve.

· 17 ·

How Natural Selection Creates Social Instincts

CHARLES DARWIN, IN *The Expression of the Emotions in Man and Animals* (1873), was the first to advance the idea that instinct evolves by natural selection. Simple in style and profusely illustrated, this last and least known of his four great books argued that the behavioral traits defining each species, no less than those defining traits of their anatomy and physiology, are hereditary. They arose and exist today, Darwin said, because in the past they aided survival and reproduction.

Darwin's fundamental insight has been verified over and again. It underpins much of what we understand about behavior today. Its potency is the reason that a century later Konrad Lorenz, one of the founders of modern animal behavior research, called Darwin the patron saint of psychology.

Yet—no idea of modern science stirred more controversy than that of human instinct as a product of mutation and natural selection. In the 1950s it survived the onslaught of radical behaviorism of the kind masterminded by B. F. Skinner, the idea that all behavior in both animals and humans is somehow and at some stage or other of each individual's development the product of learning. In the two decades that followed, the idea of instinct shaped by natural selection defeated this perception of the brain as a blank slate. At least it did so for animals. For two more decades, however, the blank slate was kept alive for human social behavior. Many writers in the social sciences and humanities continued to insist that the mind is entirely the product of its environment and past history. Free will exists and is powerful, they said. The mind is ultimately at the command of will and fate. What evolves in the mind, they finally argued, is exclusively cultural; there is no such thing as a genetically based human nature.

In fact, the evidence for instinct and human nature was already compelling at that time. Today it is overwhelming in amount and rigor, with new evidence added whenever it is tested. Instinct and human nature are increasingly the subject of studies in genetics, neuroscience, anthropology, and, nowadays, even in the social sciences and humanities themselves.

How does instinct evolve by natural selection? To keep the matter as elementary as possible, consider an imaginary population of birds nesting in a forest of mixed oak and pine. The birds choose only oak trees for their abode, a hereditary predisposition prescribed in the simplest possible way by one allele, in other words one form of two or more versions of one particular gene. Let us refer to the allele as *a*. Because of the influence of allele *a*, birds are automatically drawn to oak trees when they nest, preferring them over the numerous pine trees growing in the same forest. Their brains automatically select certain features that define oak trees. The features might be the height and contour of the canopy, for example, or the look and feel of the upper branches.

In one particular forest, an environmental shift occurs. Oak trees grow scarce because of local climate change and the inroads of a new disease. Pines, better adapted to the new conditions, begin to fill in the empty spaces. In time pines become dominant in the forest. Meanwhile, a second form of the same gene, the allele *b*, appears in the birds as a mutation of the oak-prone allele *a*. Perhaps *b* is not really a new mutation. Perhaps it has always been present at very low frequencies, sustained by mutations that have occurred rarely but repeatedly in the past. Or else pine-favoring *b* was carried in by an immigrant bird that strayed into the forest from another, mostly pine-loving population living in a nearby forest.

Whatever its origin, this second allele, *b*, causes the birds carrying it to prefer nesting in pine trees instead of oak trees. In the

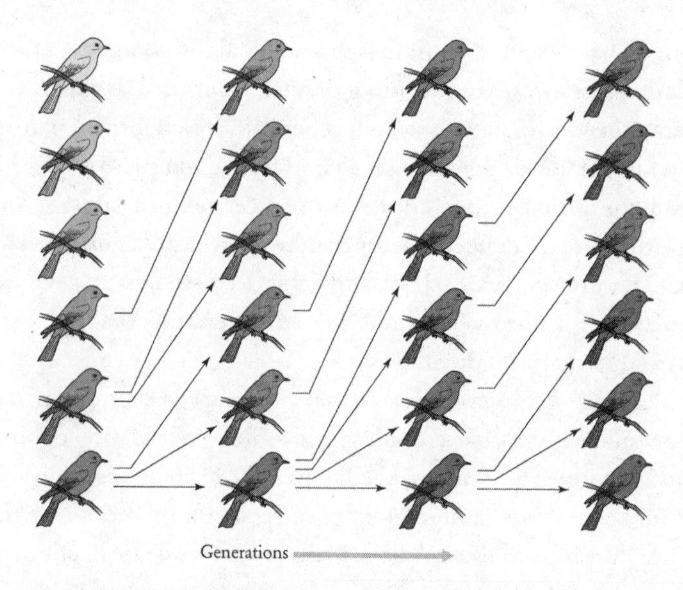

FIGURE 17-1. *Evolution by genes in its simplest form occurs when two forms (alleles) of the same gene produce different traits—in this hypothetical example, color—because of the greater survival or reproduction or both of one of the forms (darker to the right). (From Carl Zimmer,* The Tangled Bank: An Introduction to Evolution *[Greenwood Village, CO: Roberts, 2010], p. 33.)*

changing forest, where pine is rising to dominance over oak, *b* now does better than *a* or, to be a bit more precise and to the point, birds carrying *b* do better than those carrying *a*. From one generation to the next, *b* increases in frequency within the bird population as a whole. It may eventually replace *a* entirely, or not. But in either case, *evolution has occurred.* This change in the heredity of the bird population is not large compared with the rest of the birds' entire genetic code. It is an incident of "microevolution." But its consequences are great. The shift from a preponderance of allele *a* to a preponderance of allele *b* allows the bird species to continue occupying a forest now covered mostly by pine. The evolutionary change has occurred by natural selection. The changing natural environment has selected allele *b* over the previously dominant *a*. One outcome of the habitat-selection instinct has been replaced by another.

In all populations of every species, such mutations are constantly occurring in all of the traits of the species, including behavior. They may be random changes in the base pairs, the "letters" of the DNA such as the change from allele *a* to allele *b*; or a building of small portions of the DNA molecule through duplication in sequences; or changes in the number or configuration of the chromosomes that carry the DNA molecules. Most mutations harm the organism in some way or another, and as a result they soon disappear—or at best they are kept at extremely low, "mutational" levels. But a very few, like the imaginary mutant allele *b* that opened the pine forest to the previously oak-specialist birds, do provide an advantage in survival or reproductive ability, or both. As a result, they increase in frequency in the population. Additional mutations, mostly bad but a very few good, continuously appear here and there throughout the genetic code. Consequently, *evolution is always occurring.*

Although mutant alleles and other genetic novelties occur commonly over the billions of DNA letters in the vast hereditary code of billions of letters, those composing any particular gene experience such an event very rarely. One in a million or one in ten million individuals per gene each generation are typical figures. Yet if any change does occur that is favorable to survival and reproduction, as in the imagined mutation to pine-prone allele *b*, it can spread rapidly. For example, it can increase from 10 percent to 90 percent of any of the alleles in the population in as few as ten generations—even when the advantage it confers is only slight.

A vast scientific literature now exists on the dynamics of evolution, based on a century of mathematical theory coupled with empirical studies in the field and laboratory. Present-day evolutionary biology, building upon this knowledge, is growing in compass, sophistication, and power. Researchers are advancing along a wide frontier of phenomena, including sexual and asexual reproduction and the molecular foundations of particulate heredity. Scientists are also working out the interactions of multiple genes during development of the cell and organism, together with the impact of different kinds of environmental pressures on microevolution.

Taken to fine detail, the subject of evolution at the level of the gene can become forbiddingly technical. Nevertheless, several overarching principles can be gleaned that are both easily grasped and crucial for understanding the genetic basis of instinct and social behavior.

One of the principles is the distinction between the unit of heredity, as opposed to the target of selection in the process that drives evolution. The *unit* is a gene or arrangement of genes that form part of the hereditary code (thus, *a* and *b* in the forest birds). The *target* of selection is the trait or combination of traits encoded by the units of heredity and favored or disfavored by the environment. Examples of targets are propensity for hypertension and resistance to a disease in humans, or, in the case of bird behavior, the instinctive choice of nest site.

Natural selection is usually *multilevel*: it acts on genes that prescribe targets at more than one level of biological organization, such as cell and organism, or organism and colony. An extreme example of multilevel selection exists in cancer. The cancerous cell is a mutant able to grow and multiply out of control at the expense of the organism, which is the community of cells forming the next higher level of biological organization. Selection occurring at one level, the cell, can work in the opposite direction from that of the adjacent level, the organism. The runaway cancer cells cause the larger community of cells (the organism) of which it is a member, to sicken and die. Conversely, the community stays healthy when the growth of the cancer cells is controlled.

In colonies composed of authentically cooperating individuals, as in human societies, and not just robotic extensions of the mother's genome, as in eusocial insects, selection among genetically diverse individual members promotes selfish behavior. On the other hand, selection between groups of humans typically promotes altruism among members of the colony. Cheaters may win within the colony, variously acquiring a larger share of resources, avoiding dangerous tasks, or breaking rules; but colonies of cheaters lose to colonies of cooperators. How tightly organized and regulated a colony is depends on the number of cooperators as opposed to cheaters, which in turn depends on both the history of the species and the relative intensities of individual selection versus group selection that have occurred.

Traits (targets) that are acted upon exclusively by selection between groups are those emerging from interactions among members of each group. These interactions include communication, division of labor, dominance, and cooperation in performing communal tasks. If the quality of these interactions favors the colony using them over colonies using other or lesser interactions, the genes prescribing their performances will spread through the population of colonies with the passing of each generation of colonies.

Individual-versus-group selection results in a mix of altruism and selfishness, of virtue and sin, among the members of a society. If one colony member devotes its life to service over marriage, the individual is of benefit to the society, even though it does not have personal offspring. A soldier going into battle will benefit his country, but he runs a higher risk of death than one who does not. An altruist benefits the group, but a layabout or coward who saves his own energy and reduces his bodily risk passes the resulting social cost to others.

A second biological phenomenon essential to understanding the evolution of advanced social behavior is *phenotypic plasticity*. Consider a phenotype, defined as some trait of an organism prescribed at least in part by its genes. To return to the earlier imaginary example, the phenotype is the tendency of a bird to nest in either oak trees or pine trees. Next consider its genotype, the genes that prescribe the tendency to choose oak or pine trees, in this case the aforementioned alleles *a* or *b*. A phenotype prescribed by a particular genotype can be rigid in expression, such as five fingers on the hand or the color of an eye. Alternatively, it can be flexible, with its precise

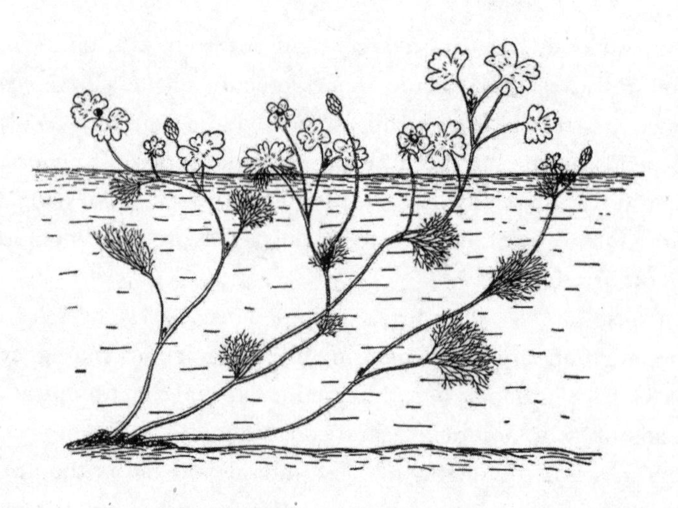

FIGURE 17-2. *The water crowfoot (*Ranunculus aquaticus*) has extreme phenotype plasticity, with leaf form determined by the location of the leaf. (From Theodosius Dobzhansky,* Evolution, Genetics, and Man *[New York: Wiley, 1955].)*

expression dependent in a predictable manner on the environment in which an individual develops. The *b* allele may prescribe a tendency to choose pine trees, but under a few conditions—perhaps rare—it chooses oak trees instead.

What is not widely appreciated, even among some biologists, is that the degree to which the amount of phenotype plasticity itself is subject to natural selection. In a classic example, the same genotype of the water crowfoot can grow one or the other of two types of leaves depending on which plant (or part of the plant) grows: broad, lobed leaves above the surface of the water, and brush-shaped leaves if under water. Both types can be produced by the same plant. And if a leaf emerges right at the water surface, the part above is broad and the part below is brush-shaped.

Finally, when thinking about evolution by natural selection, a crucial and necessary distinction to make is between *proximate causation,* which is how a structure or process works, and *ultimate causation,* which is why the structure or process exists in the first place.

Consider the imaginary forest birds as they switch from oak trees to pine trees as the place to build their nests. The proximate cause of their evolution is the possession of the *b* allele that predisposes them to choose pine over oak. More precisely, the *b* allele prescribes the development of the endocrine and nervous systems that mediate their change in nesting behavior from oak to pine. The ultimate cause is a selection pressure imposed by the environment: the decline of oak trees and their replacement by pine trees, gives the mutant allele *b* an advantage over the originally prevailing *a* allele. It is the process of natural selection that causes the population as a whole to change from allele *a* to allele *b*.

It is easy to confuse proximate and ultimate causation in particular cases, and especially in the complex multilevel process of human evolution. We frequently read, for example, that the evolutionary increase in human intelligence was caused by the invention of controlled fire, or the change to bipedal locomotion, or the employment of persistence hunting, and so forth, alone or in combinations. These innovations were landmarks in human evolution, sure enough, but not prime movers. They were preliminary steps on the pathway to the origin of the present-day high quality of human social behavior. Like the persistent nests and progressive provisioning that brought a few evolving insect species to within reach of eusociality, each step was an adaptation in its own right, with its own ultimate and proximate causes. The final step was the formation of the modern *Homo sapiens* brain, which produced the creative explosion we continue today.

· 18 ·

The Forces of Social Evolution

THE LEVEL OF biological organization at which natural selection works is a matter of profound importance in the evolution of social behavior. Does it target individuals in some way that causes their descendants to gather in groups and cooperate altruistically, because it is of such great advantage to belong to such groups? Or do kin recognize one another and form altruistic groups, because relatives share the same genes and can still place those genes in the next generation, even if they fail to do so by having offspring of their own? Or, finally, is it that hereditary altruists form groups so cooperative and well-organized as to outcompete nonaltruist groups?

The answer, supplied recently by substantial evidence, points to the last (third) explanation—in other words, group selection. To explain why this is so, I've chosen, as in the earlier chapter on the origin of social insects ("Insects Take the Giant Leap"), a mode of explanation often used in scientific publications but in this case simplified to serve a much broader public readership. The reason is that for many years I have conducted research in this field and most recently on a portion of the basic theory that has become the subject of heated controversy. The account to follow can be considered a dispatch from the scientific front.

For four decades prior to the shift to group selection, the standard explanation of ultimate causation in the evolution of advanced social behavior was inclusive-fitness theory, also called kin selection theory. Inclusive-fitness theory holds that kinship plays a central role in the origin of social behavior. In essence, it says that the more closely related individuals in a group are, the more likely they are to be altruistic and cooperative, hence the more likely are the species that formed such groups to evolve into eusociality. This notion has a powerful intuitive appeal. Why should not both ants and people favor relatives and tend to form groups united by pedigree?

For more than four decades inclusive-fitness theory had a deep effect on the interpretation of genetic evolution of all forms of social behavior. It was especially prominent as a means of addressing collateral altruism, in which individuals surrender some of their proportionate contribution to the next breeding generation to group members other than their own personal offspring.

Inclusive fitness is a product of kin selection, the means by which an individual influences the reproduction of its collateral relations, such as siblings and cousins. In a strictly biological sense, the individual is altruistic in its influence when the collateral relatives gain in genetic fitness and the altruist loses in genetic fitness. The "inclusive fitness" of the individual is its personal fitness, in other words the number of its personal offspring who grow up and have children of their own, added to the effect its actions will have on the fitness of its collateral relatives, such as siblings, aunts, uncles, and cousins. When the individual's own inclusive fitness and the fitnesses (however reduced) of its group overall increase, the gene for altruism will, according to the theory, also increase in the species as a whole. The idea of kin selection was attractive to scientists and the public from the start, valued for its apparent simplicity and the confirmation it seemed to provide for the importance of altruism in social life.

Although the idea of kin selection was first stated by the British biologist J. B. S. Haldane in 1955, the foundation of a full theory was laid out by his younger countryman William D. Hamilton in 1964. The primary formula, in what was to become the "$e = mc^2$ of sociobiology," was stated by Hamilton as an inequality, $rb > c$, meaning that an allele prescribing altruism will increase in frequency in a population if the benefit, b, to the recipient of the altruism, times r, the degree of kinship to the altruist, is greater than the cost to the altruist. The parameter r as originally expressed by Haldane and Hamilton is the fraction of genes shared by the altruist and the recipient as a result of common descent. For example, altruism will evolve if the benefit to a brother or sister is 2 times the cost to the altruist ($r = 1/2$) or 8 times to a first cousin ($r = 1/8$). To express this idea with a crude example, you will promote the altruistic gene in you if you altruistically have no children, but if your sister more than doubles the number she has as a result of your altruism to her.

No one has stated the idea of kin selection with greater clarity than Haldane in his original formulation:

> Let us suppose that you carry a rare gene which affects your behaviour so that you jump into a flooded river and save a child, but you have one chance in ten of being drowned, while I do not possess the gene, and stand on the bank and watch the child drown. If the child is your own child or your brother or sister, there is an even chance that the child will also have the gene, so five such genes will be saved in children for one lost in an adult. If you save a grandchild or nephew the advantage is only two and a half to one. If you only save a first cousin, the effect is very slight. If you try to save your first cousin once removed the population is more likely to lose this valuable gene than to gain it. But on the two occasions when I have pulled possibly drowning people out of the water (at an infinitesimal risk to myself) I had no time to make such calculations. Palaeolithic men did not make them. It is clear that genes making for conduct of this kind would only have a chance of spreading in rather small populations where most of the children were fairly near relatives of the man who risked his life. It is not easy to see how, except in small populations, such genes could have been established. Of course the conditions are even better in a community such as a beehive or ants' nest, whose members are all literally brothers and sisters.

When I first encountered the idea of kin selection in Hamilton's 1964 paper the year after its publication, I was at first skeptical. Given the enormous variety of social organizations in insect societies and our contemporary ignorance at the time of how it all came into existence, I doubted that such complexity could be fitted to such an ultrasimple equation as the Hamilton inequality. I also found it hard to believe that a newcomer in the field, and at the young age (for an evolutionary biologist) of twenty-eight, could hit upon a revolutionary new approach. (In this emotional response I overlooked my own relatively tender age of thirty-five.) After a close study, however, I changed my mind. I became enchanted by the originality and promised explanatory power of kin selection. In 1965, with Bill Hamilton at my side, I defended the idea before a mostly hostile audience at the Royal Entomological Society of London.

Hamilton was confident about the soundness of his work at that time, but depressed: his kin selection article had been rejected as a Ph.D. thesis. We walked the streets of London while I tried to buck him up. I assured him I was certain that upon resubmission the thesis would be successful, that it would have an important impact on our field. I was correct on both counts. I returned to Harvard, and in later years gave kin selection and inclusive fitness a prominent place in *The Insect Societies* (1971), *Sociobiology: The New Synthesis* (1975), and *On Human Nature* (1978), the three books that organized knowledge of social behavior into the new discipline based on population biology that I named sociobiology, and which later gave rise to evolutionary psychology. It was not, however, the Hamilton inequality itself in its abstract form that inspired me in the 1960s and 1970s. Rather, it was a brilliant suggestion by Hamilton, later to be called the haplodiploid hypothesis, that initially gave the formula its magnetic power. Haplodiploidy is the sex-determining mechanism in which fertilized eggs become females, and unfertilized eggs males. As a result, sisters are more closely related to one another ($r = 3/4$, meaning three-fourths of their genes are identical due to common descent) than daughters are to their mothers ($r = 1/2$, with half the genes identical due to common descent). Haplodiploidy happens to be the method of sex determination in the Hymenoptera, the taxonomic order comprising ants, bees, and wasps. Therefore, Hamilton said, colonies of altruistic sisters might be expected to evolve more frequently in this order than in other taxonomic orders that use conventional diplodiploid sex determination.

In the 1960s and 1970s, almost all the species known to have evolved eusociality were in the Hymenoptera. Thus the haplodiploid hypothesis seemingly had powerful support. The belief that haplodiploidy and eusociality are causally linked became standard in general reviews and textbooks of the 1970s and 1980s. The perception seemed Newtonian in concept, traveling in logical steps from an individual biological principle to a major evolutionary outcome, the pattern of occurrence of eusociality. It lent credence to a superstructure of sociobiological theory based on the presumed key role of kinship.

By the 1990s, however, the haplodiploid hypothesis began to fail. The termites had never fitted this model of explanation. Then, more eusocial groups of species came to light that were diplodiploid rather than haplodiploid in sex determination. They included one species of platypodid ambrosia beetles, several independently evolved lines of synalpheid sponge-dwelling shrimp, and two independently evolved lines of bathyergid mole rats. The result was that the connection between haplodiploidy and eusociality fell below statistical significance. Consequently the haplodiploid hypothesis has now been generally abandoned by researchers on social insects.

Meanwhile, additional kinds of evidence accumulated that proved unfavorable to the basic assumptions of kin selection and inclusive-fitness theory. One is the simple rarity of eusociality, despite the abundance of its presumed predisposition throughout the history of the animal kingdom. Vast numbers of independently evolving species are haplodiploid or clonal, the latter yielding the highest possible degree of pedigree relatedness ($r = 1$), yet without a single known case of eusociality.

It also turned out that countervailing selection forces exist that tend to make close kinship antagonistic to the evolution of altruism. They include greater genetic variability favored by group selection, as documented in the ants *Pogonomyrmex occidentalis* and *Acromyrmex echinatior*, owing, at least in the latter, to disease resistance. They also include genetic variability in predisposition to worker subcastes in *Pogonomyrmex badius*, which may sharpen division of labor and improve colony fitness—although the latter possibility has not yet been tested. Further, an increase in temperature stability of nest temperature with genetic diversity has been found within nests of honeybees and *Formica* ants. Other factors possibly working against the advantage of close pedigree kinship are the disruptive impact within colonies of nepotism, and the overall negative effects associated with inbreeding of the kind that would otherwise maximize genetic relatedness among colony members.

Most of the countervailing forces evolve through group selection or, more precisely in the case of the eusocial insects, through between-colony selection. To repeat, this level of selection is the next level above individual-level selection. It acts upon genetically based traits created by the interaction of members of a group, in particular caste determination, division of labor, communication, and communal construction of nests. The group is sufficiently well defined to reproduce itself as a unit and thereby to compete with solitary individuals and other groups of the same species.

It might seem that in theory at least the various countervailing forces in eusocial evolution can be folded into *b*, the benefit of each trait in individual fitness, and *c*, its cost, thus conserving the Hamilton inequality. In practice, however, doing so would demand a full accounting of inclusive fitness, including measures of *b* and *c*. That in turn would require field and laboratory studies of extraordinary difficulty. Nothing of this kind has been achieved, nor to my knowledge even undertaken. Further, there are mathematical difficulties with the definition of *r*, the degree of relatedness. These difficulties render incorrect the oft-repeated claim that group selection is the same as kin selection expressed through inclusive fitness.

Most writers on the subject, including its widely read champion Richard Dawkins, remained faithful, but beginning in the early

1990s, I began to have doubts. I thought it past time to ask, What did inclusive-fitness theory achieve in the explanation of altruism and altruism-based societies during three decades as the reigning paradigm of genetic social evolution? It stimulated measures of pedigree kinship and made them routine in sociobiology. These were valuable in their own right. Researchers had used the theory to predict some cases of the perturbation of sex ratios in investment by ant colonies of new reproductives; the data are overall strong, albeit consisting largely of inequalities rather than close fits. (But, as I will describe shortly, the conclusion drawn is flawed.) Kin selection theory also led to the correct prediction of effect of pedigree kinship on dominance behavior and policing. Bees and wasps that are more closely related, it was found, fight less among themselves than do those less closely related. Yet again, the conclusion drawn, that the data point to the degrees of relatedness as the key, is not the only interpretation possible. Finally, inclusive-fitness theory has been used to predict that queens of primitively eusocial bee species mate only once. However, in this case the evidence presented did not include solitary bee species as controls, so no conclusion can yet be drawn of any kind.

The results of so long a period of intense theoretical research must by any standard be considered meager. During the same period, in contrast, empirical research on eusocial organisms, and especially the insects, flourished, revealing the rich details of caste, communication, life cycles, and other phenomena and at both the individual selection and the group selection levels. Almost none of this advance was stimulated or advanced by inclusive-fitness theory, which had evolved largely into an abstract world unto itself.

Much of the inadequacy of the theory comes from looseness in the definition of *r*, hence the very concept of kinship, in various interpretations of the Hamilton inequality. The original approach taken by inclusive-fitness theorists was to define *r* as pedigree relatedness, in other words how close members of a group are in the family tree. For example, siblings are closer than first cousins. This perfectly reasonable definition pins down the average number of genes shared by two individuals owing to common descent. It was soon recognized, however, that this definition of relatedness could not work for Hamilton's equality in the majority of real and theoretical cases. As a result, different definitions were used at various times to satisfy the particular needs of the model being developed, including those designed to equate kinship models with those on multilevel natural selection. In some circumstances, kinship could be the common possession of a single allele, whether derived by pedigree or not—or even by independent mutations.

In short, the only unifying theme seemed in time to be that *r*, originally defined by pedigree, is whatever it takes to make Hamilton's inequality work. The inequality thereby lost meaning as a theoretical concept, and became all but useless as a tool for designing experiments or analyzing comparative data. In a simple model of a tag-based cooperation, for example, it turns out that the calculation of *r* involves triplet correlations. You have to pick three individuals at random from within a group, choose one as a cooperator, and the second two with the same phenotypic tag such as the same appearance or behavior (often referred to metaphorically as a "green beard"). Most biologists who knew inclusive-fitness theory only from a distance were surprised to learn that when measures are actually calculated there is no consistent biological concept behind the "relatedness" parameter.

In essence, many models have been proposed that are solved using a natural-selection, game-theoretic approach based on the idea that reproduction is proportional to payoff. It can be shown that natural selection is usually multilevel at least to some degree: its consequences at the level of the primary target trait reverberate up and down to other levels of biological organization, from molecule to population. Many of the natural-selection, game-theoretic models could be and were rephrased in terms of kin selection. To repeat, this approach, instead of looking at the direct fitness of individuals, takes in the effects of the individual's action on itself and all individuals in the group, weighted by how "related" the actor is to each recipient in turn.

It can be shown that there is a very simple resolution to this problem of diverse calculations. A general statement of dynamical natural selection is set up, then an attempt made to interpret it both ways. When this is done, it turns out that the interpretation by standard natural selection is appropriate for all cases, whereas the interpretation by kin selection, although possible in a very few cases, cannot be generalized to cover all situations without stretching the concept of "relatedness" to the point where it loses meaning.

It has become clear from a fuller foundational analysis that the Hamilton inequality permits cooperators within a group to be more than marginally abundant only under stringently narrow conditions. And it does not give a description of the underlying evolutionary dynamics, in which conditions are specified for a stationary distribution in evolution.

An important concept needed to evaluate the limitation of kin selection in real populations is weak selection. The game played by competing genotypes includes selection that might arise from response based on relatedness plus that based on every other hereditary difference among individuals, thence on all the individuals throughout everything that happens to the individual and its responses throughout its life. If two individuals are very close to one another in relatedness, they can experience some kin selection—if in fact it exists—but then the closeness damps variation in the rest of the genome among individuals, spreads the selection force over the variation that does exist, and hence reduces the amount of dynamical evolution possible. Under certain assumptions and for weak selection, the inclusive-fitness approach and the multilevel-selection approach are identical. However, as one moves away from weak selection or if the assumptions are not fulfilled, the kin selection approach cannot be generalized further without making it so broad and abstract as to lose meaning. With this perception in mind, it makes sense to ask the following question. If there is a general theory that works for everything (multilevel natural selection) and a theory that works only for some cases (kin selection), and in the few cases where the latter works it agrees with the general theory of multilevel selection, why not simply stay with the general theory everywhere?

Worse, unwarranted faith in the central role of kinship in social evolution has led to the reversal of the usual order in which biological research is conducted. The proven best way in evolutionary biology, as in most of science, is to define a problem arising during empirical research, then select or devise the theory that is needed to solve it. Almost all research in inclusive-fitness theory has been the opposite: hypothesize the key roles of kinship and kin selection, then look for evidence to test that hypothesis.

The most basic flaw in this approach is that it fails to consider multiple competing hypotheses. When biological details of particular cases are examined before inclusive-fitness theory is applied, such alternative examinations come quickly to attention. Even in the most meticulously analyzed cases presented by various authors as evidence for kin selection, it has been easy to devise explanations from standard natural-selection theory that are at least equally valid. They entail straightforward individual or group selection, or both. Kin selection may occur, but there is no case that presents compelling explanation for its role as the driving force of evolution.

A classic example to prove the need for multiple competing hypotheses is provided by the microbial biofilms and stalk-forming cellular slime molds. Free-living single-celled organisms either form mats (the case in bacteria) or else are attracted to others of the same genetic strain to form dense aggregates (slime molds). Many then take positions that reduce or sacrifice their own reproduction—clearly for the good of the group. Inclusive-fitness theoreticians have suggested that kin selection is the driving force behind this altruism. However, group selection overcoming "selfish" individual selection appears to be the more straightforward and comprehensive explanation.

A comparable interplay of multilevel-selection forces becomes evident upon close examination of the number of times eusocial ants, bees, and wasps mate. One team of inclusive-fitness theorists found that species possessing relatively primitive social organization mate with only one male and thus produce closely related offspring. The authors present their data as correlative evidence of kin selection. However, comparable data were not provided for solitary species closely related to the eusocial examples; hence there were no controls for the conclusion that single mating favors the origin of eusocial behavior. In fact, it is logical to suppose that such queens of solitary species also mate with one male only, and for a reason unrelated to kin selection: prolonged mating excursions increase the risk to young females from predators. Of equal importance, the inclusive-fitness researchers pointed to the origin of multiple-male matings practiced by queens of many of the hymenopteran species with advanced colonial organization. This, they concluded, indicates the relaxation of kin selection in later stages of evolution. But they overlooked the near-limitation of multiple-male mating to species with exceptionally large worker populations, shown in their own data. Here, group selection favoring stored sperm or resistance to pathogen threat in large nests, or both, is more plausibly the driving force.

A second class of explanations for the origin of advanced social behavior that emerges from case-by-case assessments using standard natural-selection theory is discordance among the group members as a factor in the evolution of physiology and behavior. The more distantly related the members, the less likely they are to communicate effectively, to respond to the same cues in the environment, and to coordinate their activities with precision. A genetically very diverse group is prone to be less harmonious and hence eliminated by group selection. The same principle applies to an extreme in the more familiar cases of cancer cells in an organism and, at another level of biological organization, to the genetic isolating mechanisms that divide single species into two or more daughter species. Further, interplay of individual selection and group selection in microbial societies can be viewed as suppressing discordance of the participant cells. In this interpretation, an alternative to that implied by inclusive fitness, successfully cooperating cells are plastic variants of the same genotype, and colony formation is the result of group selection that works against discordance from mutant phenotypes.

The same basic argument applies to the role of nutrition in the control of queen production of honeybees, in which workers feed larvae a special food, royal jelly, that turns them into queens. It is also relevant to restraint and policing in the control of worker reproduction in insect societies generally. Both classes of phenomena have been framed at times in the language of kin selection and its product inclusive fitness, but reduction of discordance by group selection with no kin selection is at least equally plausible.

A stanchion of inclusive-fitness theory has long been the explanation of how and why ant colonies regulate the amount of food they invest in the production of virgin queens versus males. If the mother was singly mated, she should in theory wish a ratio of one male to one female, since she is equally related (half of the group share genes by common descent) to her daughters, the virgin queens, and to her sons, the reproductive males. However, as argued by Robert L. Trivers and Hope Hare in 1976 and elaborated at great length by inclusive-fitness theorists with species of ants, the workers should wish more investment in virgin queens, their sisters, since they share three-fourths of their genes by common descent, because of the haplodiploid mode of sex determination. In contrast, they share only one-fourth of their genes with the males, their brothers. Therefore, the argument goes, the mother queen and her worker daughters are in conflict over the sex ratio of new reproductives produced by the colony. Many studies have in fact shown that the royal ratio is slanted in favor of producing queens. The workers thus appear to have won the conflict, and inclusive-fitness theory is confirmed.

The inclusive-fitness approach to reproductive sex ratio determination in ants is one of the most elaborated and documented bodies of theory in evolutionary biology. Yet it is based on two starting assumptions, that pedigree relatedness is a primary determining factor of sex ratio, and, following from this first assumption, that groups within the colony with different group-level degrees of relatedness are in conflict. What if one, or both, of these assumptions were not correct? A simpler and more straightforward explanation is available from elementary natural-selection theory, in the absence of kin selection, as follows. The goal of the whole colony is to put as many future parents into the next generation as possible. In ant species generally, males are smaller and lighter than virgin queens, often strikingly so, because of the heavy fat reserves the queens must carry in order to start new colonies. Males cost less to make, and if the ratio of energy investment were 1:1, more males than queens would be available for mating. Most commonly the young reproductives have only one chance to mate, so that, on average, producing an excess of males would be a waste for the colony. Only if the colony had knowledge of perturbations of production ratios of other colonies, or the mortality of males in the nuptial flights were greater, could it choose otherwise. As a result, it is in the best interest of both the mother queen and her worker daughters to bias energy investment in favor of virgin queens. This explanation, freed from the assumptions of kin selection, and with colony-level selection added, is more consistent with the data than the explanation from inclusive-fitness theory. In species with multiple mother queens and in slave-making colonies, virgin queens typically do not need the heavy body reserves to found colonies independently, and hence, as occurs in nature, the ideal ratio is predicted to be closer to 1:1. These trends are also consistent with the data. Further perturbation of sex ratios apparently reflects selection pressures from the particular environments in which colonies either launch their virgin queens and males on mating flights or else keep them home until they mate.

In another, very different setting, a similarly meticulous experimental analysis has demonstrated that in the periodically subsocial eresid spider *Stegodyphus lineatus*, groups of sibling spiderlings extract more nutrients from communal prey than do spiderling groups of artificially mixed parentage. Because the researchers believe that spiderlings withhold injecting digestive enzymes in order to avoid exploitation by strangers, they accept the kin selection hypothesis. Yet a quick calculation shows that such behavior would reduce the average payout for each individual, including those that withhold their digestive enzymes. The reduction in communal intake could be better explained either by discordance in cues among unrelated spiderlings or by overt conflict among them.

The expectation of inheritance is a third process that can lead to seeming kin-based altruism but is more simply and realistically explained as the straightforward result of individual-level selection. In a small percentage of bird and mammal species, offspring remain at the nest of their birth and assist their parents in rearing additional broods. They thereby delay reproduction on their own while increasing reproduction of their parents. Inclusive-fitness researchers have attributed the phenomenon to kin selection, and bolstered their argument by demonstrating a positive correlation across species between closeness of kinship and the amount of help provided to parents by the stay-at-homes. However, more thorough, previously published studies covering a wide range of species life history data had already arrived at a different explanation, entailing multilevel selection with a strong weight placed on individual-level selection. Under certain conditions unrelated to kin selection, the persistence of adult young at the natal nest is favored. The conditions include unusual scarcity either of nest site or territory or both, or alternatively, low adult mortality or relatively unchanging conditions in a stable environment. After prolonged residence, the helpers inherit the nest or territory upon the death of the parents. The positive correlation across species between kinship and helping reported by the inclusive-fitness researchers is based on only a few data points and can be logically explained by the common practice of a "floating strategy" in some species, in which individuals move about nests and spread the amount of help given. The more the floating, the less the average kinship and help given at each nest visited.

I was able personally to examine the helper phenomenon in the red-cockaded woodpecker when I visited a population in West Florida and discussed the details with researchers who had followed the personal life histories of birds tagged for identification in the wild. The red-cockaded is the only woodpecker species in the world, I learned, that digs its nests in the trunks of living trees. It takes a young male as much as a year to build such a nest, and the location must also be outside the territories of established families. Until then, it is to the advantage of both daughters and sons to stay home. Further, during the waiting period, one or both parents may die, and the natal nest can be inherited. It is moreover to the advantage of the parents to tolerate grown children only if they work as helpers.

The essential line of reasoning in inclusive-fitness theory, to summarize, has been as follows. Kin selection is assumed to occur and to be in fact inevitable in many biological systems. When kin selection occurs, it is following the Hamilton inequality, which predicts in the simplest case at least whether genes for altruism will increase in the population at large or not. When Hamilton's inequality is applied to all the members of a group, it yields the inclusive fitness for the group, which, if known, can predict whether a population of such groups is evolving toward an altruism-based social organization.

None of these assumptions, however, has been found to hold up. Empiricists who have measured genetic relatedness and use inclusive-fitness arguments have thought that they were placing their reasoning on a solid theoretical foundation. Such is not the case, however. Inclusive fitness is a special mathematical approach with so many limitations as to make it inoperable. It is not a general evolutionary theory as widely believed, and it characterizes neither the dynamics of evolution nor the distributions of gene frequencies.

In the extreme cases where inclusive-fitness theory might work, biological conditions are required that demonstrably do not exist in nature. The system, it turns out, must move to the mathematical limit of "weak selection," in which all members of a group approach the same fitness, and all alternative responses must be about equally abundant. Further, all interactions among the colony members must be additive and pairwise, one on one. In fact, all known societies other than mated pairs violate this condition. Other kinds of interactions tend to be synergistic to a degree that varies with the constantly changing condition of the colony. Finally, inclusive-fitness theory can be used only in static structures in which the intensities of interaction cannot vary from one contact to another, and there must be a global updating in cycles.

This issue of theoretical biology is important, because the intuition provided by inclusive-fitness theory has been widely if mistakenly embraced as generally correct. In fact, inclusive-fitness arguments without fully specified models, of the kind ordinarily advanced by field and laboratory researchers, are misleading. How far off the reasoning can be is illustrated by the mathematical demonstration that all measures of relatedness can be identical in two systems, yet cooperation is favored in one system and not in the other. Conversely, two populations can have relatedness measures on the opposite ends of the spectrum and yet both structures be equally unable to support the evolution of cooperation.

Another commonly held misconception is that inclusive-fitness calculations are simpler than those of standard natural-selection models. That is not the case. In the rare cases where inclusive fitness can be made to work in abstract models, the two theories are identical and require the measurement of the same quantities.

The old paradigm of social evolution, grown venerable after four decades, has thus failed. Its line of reasoning, from kin selection as the process, to the Hamilton inequality condition for cooperation, and thence to inclusive fitness as the Darwinian status of colony members, does not work. Kin selection, if it occurs at all in animals, must be a weak form of selection that occurs only in special conditions easily violated. As the object of general theory, inclusive fitness is a phantom mathematical construction that cannot be fixed in any manner that conveys realistic biological meaning. Nor can it be used to track the evolutionary dynamics of genetically based social systems.

The misadventure of inclusive-fitness theory originated in the belief that a single abstract formulation, in this case the Hamilton inequality, has implications that can be unpacked layer by layer to account for social evolution in ever-growing detail. This belief can be refuted by both mathematical logic and empirical evidence. What, then, is the best direction we should take to understand advanced social behavior?

· 19 ·

The Emergence of a New Theory of Eusociality

THE EVOLUTIONARY ORIGIN of any complex biological system can be reconstructed correctly only if viewed as the culmination of a history of stages tracked from start to finish. It begins with empirically known biological phenomena in each stage, if such is known, and it explores the range of phenomena that are theoretically possible. Each transition from one stage to the next requires different models, and each needs to be placed in its own context of potential cause and effect. This is the only way to arrive at the deep meaning of advanced social evolution and the human condition itself.

The first conceivable stage in the origin of eusociality, entailing division of labor that is seemingly altruistic, is the formation of groups within a freely mixing population of otherwise solitary individuals. There are in theory many ways in which this might occur in reality. Groups can assemble when nest sites or food sources on which a species is specialized are local in distribution, or when parents and offspring stay together, or when migratory columns branch repeatedly before settling, or when flocks follow leaders to known feeding grounds. They might even come together randomly by mutual local attraction.

The way in which groups are formed probably has a profound effect on the likelihood of progress toward eusociality. The most important way includes the tightening of group cohesion and persistence. For example, as I have stressed, all of the evolutionary lines known with primitively eusocial species surviving (in aculeate wasps, halictine and xylocopine bees, sponge-nesting shrimp, termopsid termites, colonial aphids and thrips, ambrosia beetles, and naked mole rats) have colonies that build and occupy defensible nests. In a few cases, unrelated individuals join forces to create the little fortresses. Unrelated colonies of *Zootermopsis angusticollis*, for example, fuse to form a supercolony with a single royal pair through repeated combats. In most cases of animal eusociality, however, the colony is begun by a single inseminated queen (for instance in the Hymenoptera) or mated pair (termites). Therefore, in most cases the colony grows by the addition of offspring that serve as nonreproductive workers. In a few, more primitively eusocial species, the growth is hastened by the acceptance of alien workers or by the cooperation of unrelated founding queens.

Grouping by family can accelerate the spread of eusocial alleles, but it does not of itself lead to advanced social behavior. The causative agent of advanced social behavior is the advantage of a defensible nest, especially one expensive to make and within reach of a sustainable supply of food. Because of this primary condition in the insects, close genetic relatedness in primitive colony formation is the consequence, not the cause, of eusocial behavior.

The second stage is the happenstance accumulation of other traits that make the change to eusociality still more likely. The most important is close care of the growing brood in the nest—by feeding the young progressively, or cleaning the brood chambers, or guarding them, or some combination of the three. Like constructing a defensible nest by the solitary ancestor, these preadaptations arise by individual-level selection, with no anticipation of a future role in the origin of eusociality (anticipation is absent because evolution by natural selection cannot predict the future). The preadaptations are products of adaptive radiation, in which species split and spread into ecologically different niches. According to the niches on which they specialize, some of the species are more likely than others to acquire potent preadaptations. Some species, for example, may come to live in habitats relatively free of predators. Having a less urgent need to protect the brood, they are likely to remain stable in social evolution or evolve away altogether to a solitary life. Others, in habitats thick with dangerous predators, will draw closely to the threshold of eusociality and make its crossing more likely. The theory of this stage is the theory of adaptive radiation, worked out already by many researchers independently of studies on eusociality.

The third step in evolution to advanced social behavior is the origin of the eusocial alleles, whether by mutation or by immigration of mutant individuals from the outside. In preadapted hymenopterans (bees and wasps) at least, this event can occur as a single point mutation. Further, the mutation is not required to prescribe the construction of a novel behavior. It need simply cancel an old one. Crossing the threshold to eusociality requires only that a female and her adult offspring fail to disperse to start new, individual nests. Instead, they remain at the old nest. At this point, if environmental selection pressures are strong enough, the spring-loaded preadaptations kick in and members of the group commence the interactions that turn them into a eusocial colony.

Eusocial genes have not yet been identified, but at least two other genes or small ensembles of genes are known that prescribe major changes in social traits by silencing mutations in preexisting traits. These examples, and the promise they offer of advances in both theory and genetic analysis, bring us to the fourth phase in the evolution of animal eusociality. As soon as the parent and subordinate offspring remain at the nest, as with a primitively social family of bees or wasps, group selection proceeds, uniquely targeting the emergent traits created by the interactions of the colony members. The selection forces will probably create an alerting system with alarm calls or chemical signals. They will develop odors on their bodies to distinguish their colony from others. They are likely to invent the means to draw nestmates to newly discovered food. At least in the more advanced stages, they will evolve differences in anatomy and behavior between the royal reproductives and the supporting worker caste.

By looking at the emergent traits on which group selection acts, it is possible to envision a new mode of theoretical research. Among the phenomena newly highlighted is that the different roles of the reproductive parents and their nonreproductive offspring are not genetically determined. Rather, as evidence from primitively eusocial species has shown, they represent alternative phenotypes of the same genotype. In other words, the queen and her workers have the same genes that prescribe caste and division of labor, although they vary extensively in other genes. This circumstance lends credence to the view that the colony can be viewed as an individual organism or, more precisely, an individual superorganism. Further, insofar as social behavior is concerned, descent is from queen to queen, with the worker force as an extension of each in turn. Group selection still occurs, but it is conceived to be selected as the traits of the queen and the extrasomatic projection of her personal genome. This perception has opened a new form of theoretical inquiry, as well as questions that can only be settled by a new focus of empirical research.

The fourth phase is identification of the environmental forces driving group selection, which is the logical subject of combined investigations in population genetics and behavioral ecology. Research programs have scarcely begun in this area, in part because of the relative neglect of the study of the environmental selection forces that shape early eusocial evolution. The natural history of the more primitively eusocial animals, and especially structure of their nests and fierce defense of them, suggests that a key element in the origin of eusociality is defense against enemies, including parasites, predators, and rival colonies. But very few field and laboratory experimental studies have been devised to test this and potential competing hypotheses.

In the fifth and final phase, group (between-colony) selection shapes the life cycle and caste systems of the more advanced eusocial species. As a result, many evolutionary lines have evolved very specialized and elaborate social systems. The ultimate such systems are found not in humans but in insects, particularly those at the most advanced level—the honey bees, stingless bees, leafcutter ants, weaver ants, army ants, and mound-building termites.

In briefest terms, a full theory of eusocial evolution will consist of a series of stages, subject to experimental verification, of which the following may be recognized:

1. The formation of groups.
2. The occurrence of a minimum and necessary combination of preadaptive traits in the groups, causing the groups to be tightly formed. In animals at least, the combination includes a valuable and defensible nest. The nest-dependent condition predetermines the likelihood that primitively eusocial groups will be a family—parent and offspring in insects and other invertebrates, and extended families in vertebrates.
3. The appearance of mutations that prescribe the persistence of the group, most likely by the knockout of dispersal behavior. Evidently, a durable nest remains the key element in maintaining the prevalence. Primitive eusociality may emerge immediately due to spring-loaded preadaptations—those evolved in earlier stages that by chance cause groups to behave in a eusocial manner.
4. In the insects, emergent traits caused by either the genesis of robot-like workers or the interaction of group members are shaped through group-level selection by environmental forces.
5. Group-level selection drives changes in the insect colony life cycle and social structures, often to bizarre extremes, producing elaborate superorganisms.

Given that the last two steps occur only in the insects and other invertebrates, how, then, did the human species achieve its own unique, culture-based social condition? What mark has the combined genetic and cultural process put on human nature? Stated another way, *what are we*?

V

What Are We?

· 20 ·

What Is Human Nature?

SURELY ALL WILL AGREE: a clear definition of human nature is the key to understanding the human condition as a whole. But the achievement of that definition, it turns out, is an extraordinarily difficult task. Human nature is obvious through its manifestation in everyday life. Its intuitive expression is the substance of the creative arts and the underpinning of the social sciences. Yet its true identity has remained elusive. There may be an emotional, very human reason for this persistent ambiguity. If raw, untransformed human nature were to be revealed, and the philosopher's stone thus attained, what would it be? What would it look like? Would we love it? A better question may be: Do we really want to know?

Perhaps most people, including many scholars, would like to keep human nature at least partly in the dark. It is the monster in the fever swamp of public discourse. Its perception is distorted by idiosyncratic personal self-regard and expectation. Economists have by and large steered around it, while philosophers bold enough to search for it have always lost their way. Theologians tend to give up, attributing it in different parts to God and the devil. Political ideologues ranging from anarchists to fascists have defined it to their selfish advantage.

The very existence of human nature was denied during the last century by most social scientists. They clung to the dogma, in spite of mounting evidence, that all social behavior is learned and all culture is the product of history passed from one generation to the next.

Leaders of conservative religions, in contrast, have been prone to believe that human nature is a fixed property vouchsafed by God—to be explained to the masses by those privileged to understand His wishes. Paul VI, in his 1969 encyclical *Humanae Vitae,* for example, explained, "Man cannot attain that true happiness for which he yearns with all the strength of his spirit, unless he keeps the laws which the Most High God has engraved in his very nature. These laws must be wisely and lovingly observed." In particular, he said, the divine laws of human nature forbid any use of artificial contraception.

I believe that ample evidence, arising from multiple branches of learning in the sciences and humanities, allows a clear definition of human nature. But before suggesting it, let me first explain what it is not. Human nature is not the genes underlying it. They prescribe the developmental rules of the brain, sensory system, and behavior that produce human nature. Nor can the universals of culture discovered by anthropologists be defined collectively as human nature. The following, for example, are the sixty-seven social behaviors and institutions shared by all of the hundreds of societies in the Human Relations Area Files, as compiled in the classic 1945 study by George P. Murdock, and here alphabetically listed:

> age-grading, athletic sports, bodily adornment, calendar, cleanliness training, community organization, cooking, cooperative labor, cosmology, courtship, dancing, decorative art, divination, division of labor, dream interpretation, education, eschatology, ethics, ethno-botany, etiquette, faith healing, family feasting, fire-making, folklore, food taboos, funeral rites, games, gestures, gift-giving, government, greetings, hair styles, hospitality, housing, hygiene, incest taboos, inheritance rules, joking, kin groups, kinship nomenclature, language, law, luck superstitions, magic, marriage, mealtimes, medicine, obstetrics, penal sanctions, personal names, population policy, postnatal care, pregnancy usages, property rights, propitiation of supernatural beings, puberty customs, religious ritual, residence rules, sexual restrictions, soul concepts, status differentiation, surgery, tool-making, trade, visiting, weather control, and weaving.

It is tempting to suppose that this list is not only truly diagnostic for human beings but inevitable for the evolution of any species in any star system that reaches the human level of high intelligence and complex language, regardless of its undergirding hereditary predispositions. However, that is almost certainly not the case, because it is possible to imagine other worlds in which large terrestrial creatures evolve different combinations of cultural traits. It would be premature to expect each of such theoretical universals to be genetic in nature. In any case, the human universals are better seen as the predictable products of something deeper.

If the genetic code underlying human nature is too close to its molecular underpinning and the cultural universals are too far away from it, it follows that the best place to search for hereditary human nature is in between, in the rules of development prescribed by genes, through which the universals of culture are created.

Human nature is the inherited regularities of mental development common to our species. They are the "epigenetic rules," which evolved by the interaction of genetic and cultural evolution that occurred over a long period in deep prehistory. These rules are the genetic biases in the way our senses perceive the world, the symbolic coding by which we represent the world, the options we automatically open to ourselves, and the responses we find easiest and most rewarding to make. In ways that are beginning to come into focus at the physiological level and, even in a few cases, genetic level, epigenetic rules alter the way we see and linguistically classify color. They cause us to evaluate the aesthetics of artistic design according to elementary abstract shapes and the degree of complexity. They determine the individuals we as a rule find sexually most attractive. They lead us differentially to acquire fears and phobias concerning dangers in the environment, as from snakes and heights; to communicate with certain facial expressions and forms of body language; to bond with infants; to bond conjugally; and so on across a wide range of other categories in behavior and thought. Most epigenetic rules are evidently very ancient, dating back millions of years in our mammalian ancestry. Others, like the stages of linguistic development, are only hundreds of thousands of years old. At least one, adult tolerance to lactose in milk and from that the potential for a dairy-based culture in some populations, dates back only a few thousand years.

As *epi-* in the word "epigenetic" implies, the rules of physiological development are not genetically hardwired. They are not beyond conscious control, like the autonomic "behaviors" of heartbeat and breathing. They are less rigid than pure reflexes such as eyeblinks and knee jerks. The most complex of reflexes is the startle response. If you come up unseen behind another person and make a sudden loud noise—a shout, a crashing of two objects together—he will, in a fraction of a second, faster than the frontal cortex can process the response, relax his body, close his eyes, open his mouth, drop his head forward, and bend his knees slightly. In nature and modern life, the response instantly and unconsciously prepares him for the collision or blow likely to follow. His life may be saved at another time from the onslaught of an enemy or predator. The startle response is rigidly prescribed by genes, but it is not part of human nature as we intuitively perceive it. It is a typical reflex, performed entirely outside the conscious mind.

The behaviors created by epigenetic rules are not hardwired like reflexes. It is the epigenetic rules instead that are hardwired, and hence compose the true core of human nature. These behaviors are learned, but the process is what psychologists call "prepared." In prepared learning, we are innately predisposed to learn and thereby reinforce one option over another. We are "counterprepared" to make alternative choices, or even actively to avoid them. For example, we are prepared to learn a fear of snakes very quickly, proceeding easily to the point of phobia, yet we are not prepared by instinct to treat other reptiles, such as turtles and lizards, with any such degree of revulsion. We are attracted through prepared learning to find beauty in a stream-crossed parkland, and counterprepared to do the same for the interior of dark forests. Such responses seem "natural" to us, even though they must be learned, and that is precisely the point.

How are such epigenetic rules of learning evolved? I began to think a great deal about the process in the 1970s, when controversies over heredity-versus-environment and genes-versus-culture were political and at white heat. The root of the problem, as I saw it, was the manner in which the evolution of genes affects the evolution of culture. This interaction, it turned out, presented a theoretical challenge of exceptionally interesting difficulty.

In 1979 I invited Charles J. Lumsden, a young theoretical physicist of demonstrated ability, to join me in a study of this subject. We soon came to realize that the process can be unraveled only if we treat its mystery as not one but as two unsolved problems. The first problem was to identify the instinctive, hence noncultural basis of human nature. The second, even less tractable problem was the causal relation between the evolution of genes and the evolution of culture, or "gene-culture coevolution," as we decided to call it. It had been apparent for some time that many properties of human social behavior are affected by heredity, both for the species as a whole and for differences among members of the same population. It was also clear that the innate properties of human nature must have evolved as adaptations. We surmised, too, that the key to the solution is the preparedness and counterpreparedness in how people learn culture. In the following two years, Lumsden and I constructed and presented the first theory of gene-culture coevolution.

Other researchers picked up the notion of gene-culture coevolution, while however putting strong emphasis on cultural evolution. They saw genetic evolution principally as a force that has given rise to the capacity for culture, or else as one in a dual track running more or less separately alongside cultural evolution. They paid little attention to the interactions, epigenetic rules, or the genetic components by which coevolution occurs.

This one-sidedness is curious, given substantial evidence already in hand during the 1970s and 1980s of genetic properties

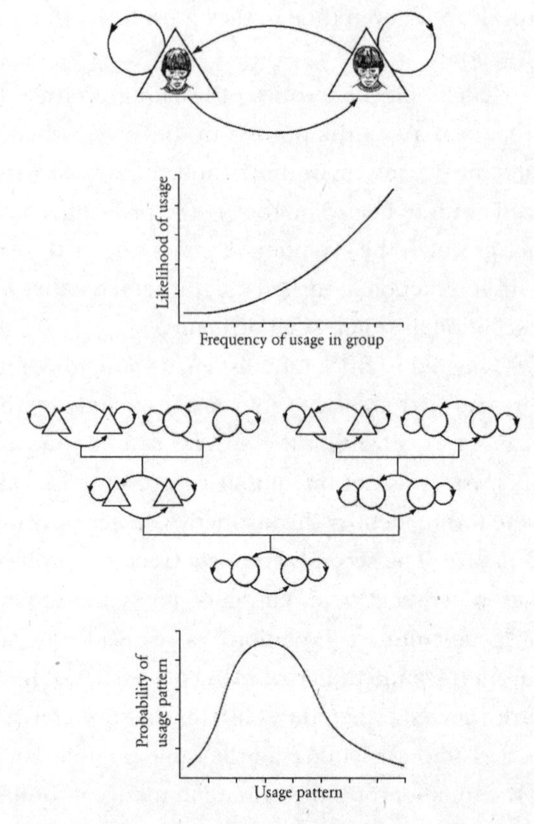

FIGURE 20-1. *The dynamics of gene-culture coevolution. The stages that lead from individual decision making to the creation of diversity among cultures is illustrated by body decoration in the Tapirapé Indians of Brazil. The processes are expressed in abstract form, following from the theory of gene-culture coevolution. Proceeding from the top down, the sequence is as follows: the individual chooses whether or not to adorn his body, and he switches from one option to the other at a certain rate; his rate of change depends on the frequency with which others express a preference for one choice or the other; each of the individuals in a tribal group (illustrated in the third panel down) or society is either using body adornment or not; from the above information, the anthropologist (bottom panel) can estimate the probability that a certain percentage in the group uses adornment, that is, a particular usage pattern exists, at a given moment in time. (From Charles J. Lumsden and Edward O. Wilson,* Promethean Fire: Reflections on the Origin of Mind *[Cambridge, MA: Harvard University Press, 1983].)*

of the kind usually cited as part of "human nature," with palpable influences on some aspects of cultural evolution. The bias may have arisen as an excess of caution in deference to the "blank-slate" view of the mind, which denied the existence of human instinct altogether. The general preference in the 1970s and 1980s favored instead what might be called the "promethean gene" hypothesis. Genetic evolution produced culture, according to supporters of this view, but only in the sense that it created the capacity for culture. Social scientists during that period, with a few notable exceptions, accepted both the blank-slate brain and the promethean gene as a way of affirming the autonomy of the social sciences and the humanities. This biologically nondimensional view of social evolution was further deduced from a second key hypothesis, the psychic unity of mankind. This opinion held that human culture evolved during too short a time for genetic evolution to have occurred, at least beyond the all-purpose promethean genotype that separates humanity from other animal species.

At first thought, it might seem that cultural evolution would indeed tend to inhibit or even reverse genetic evolution. The use of campfires, enclosed dwellings, and warm clothing allowed humans to survive and reproduce in parts of the world where survival through winter would otherwise have been impossible. Furthermore, improved methods of hunting and the planting of crops allowed people to flourish in habitats where they would normally have faced starvation. Why, it is then reasonable to ask, be ruled by genes if cultural changes could achieve the same result in such short order?

In fact, cultural evolution undoubtedly does tend to smother genetic evolution. Even so, there are novel challenges and opportunities abounding in the world's many habitats that can also be met—or at least met more effectively—by a change in genes guided by natural selection, including strange new foods, diseases, and climatic regimes. The explosion of new mutations that occurred following the breakout from Africa some 60,000 years ago created large numbers of such potentially adaptive new genes. It would be surprising that genetic evolution has not occurred in different populations as they colonized the rest of the world.

The textbook example of gene-culture coevolution occurring in recent millennia is the development of lactose tolerance in adults. In all previous human generations, the production of lactase, the enzyme that converts the sugar lactose into digestible sugars, was present only in infants. When children were weaned off their mother's milk, their bodies automatically shut down further production of lactase. When herding was developed 9,000 to 3,000 years ago, variously and independently in northern Europe and East Africa, mutations spread culturally that sustained lactase production into adult life, allowing the continued consumption of milk. The advantage to survival and reproduction in utilizing milk and milk products proved enormous. Herds of dairy cows, goats, and camels are among the most productive and reliable year-round sources of food available to humans. Four independent mutations have been discovered by geneticists that prolong lactase production, one in Europe and three in Africa.

Lactose tolerance is an example of what ecologists and researchers on human evolution call "niche construction." In the case of gene-culture coevolution of lactose production, the niche was created to include cattle domestication as a major new source of food. Mutant genes were available in very low frequencies, and they rapidly replaced the other, older variants. They were moreover protein-coding genes, the principal means by which changes occur in specific tissues, in this case the alimentary canal.

Over the past half century, large numbers of other such intertwined coevolutionary processes have been uncovered by anthropologists and psychologists. Put together, they form a class of genetic changes different in kind from the local acquisition of lactose tolerance. They are universal in modern humanity and also ancient, their origins predating the emergence of modern *Homo sapiens* and at least in some cases even the human-chimpanzee split of more than six million years ago. Working at the level of cognition and emotion, their effect on the evolution of language and culture has been both deep and wide. They make up much of what is intuitively called "human nature."

One of the most important and best understood examples is incest avoidance. Incest taboos are a cultural universal. All of the hundreds of societies that have been studied by anthropologists tolerate and occasionally even encourage marriage between first cousins but forbid it between siblings and half siblings. A very few societies in historical times have institutionalized brother-sister incest for some of its members. The roster includes Incas, Hawaiians, some Thais, ancient Egyptians, Monomotapa of Zimbabwe, Ankale, Buganda, and Bunyoro of Uganda, Nyanza of the Congo, Zande and Shilluk of Sudan, and Dahomeans. In each case the practice was surrounded by ritual and limited to royalty or other groups of high status. Political power was transmitted through the male line, and multiple wives were permitted to the men, allowing them to father separate, nonincestuous children.

Elsewhere brother-sister incest is strictly avoided. A personal revulsion against it is socially reinforced in most cultures by taboo and law. The risk of having defective children through incest is well understood. On average, each person carries somewhere on his twenty-three pairs of chromosomes at least two sites bearing recessive genes that are defective to some degree, and in extreme cases lethal. At each site, the recessive gene occurs on one chromosome, and its counterpart on the other is normal. When both chromosomes carry the defective gene, the person carrying them develops the disease—or at least a greater likelihood of acquiring it. The defect can occur even in the womb, resulting in a spontaneous abortion. If, on the other hand, one of the two genes is normal, it overrides the impact of the defective gene, and the individual develops normally. Hence the term "recessive": the gene is hidden in the presence of its normal, "dominant" counterpart. The vulnerable sites are now known to include both protein-coding genes and regulatory regions of the

DNA between the genes. Such diseases, either outright recessive or mostly recessive in genetic control, include macular degeneration, inflammatory bowel disease, prostate cancer, obesity, type 2 diabetes, and congenital heart disease.

The destructive consequence of incest is a general phenomenon not just in humans but also in plants and animals. Almost all species vulnerable to moderate or severe inbreeding depression use some biologically programmed method to avoid incest. Among the apes, monkeys, and other nonhuman primates, the method is two-layered. First, among all nineteen social species whose mating patterns have been studied, young individuals tend to practice the equivalent of human exogamy. Before reaching full adult size, they leave the group in which they were born and join another. In the lemurs of Madagascar and in the majority of monkey species from both the Old and the New Worlds, it is the males who emigrate. In red colobus monkeys, hamadryas baboons, gorillas, and chimpanzees of Africa, the females leave. In howler monkeys of Central and South America, both sexes depart. The restless young of these diverse primate species are not driven out of the group by aggressive adults. Rather, their departure appears to be entirely voluntary.

In humans, precisely the same phenomenon occurs in the form of exogamy, in which young adults, usually women, are exchanged between tribes. The consequences of exogamous exchanges in culture are many, and have been analyzed in detail by anthropologists. For the explanation of the origin of exogamy as an instinct of profound genetic value, however, one need look no further than the universal pattern followed by all other primate species.

Whatever its ultimate evolutionary origin, and however else it affects reproductive success, the emigration of young primates prior to reaching full sexual maturity greatly reduces the potential for inbreeding. But the barrier against inbreeding is reinforced by a second line of resistance. This is the avoidance of sexual activity among closely related individuals who remain with their natal group. In all the social nonhuman primate species whose sexual development has been carefully studied, including marmosets and tamarins of South America, Asian macaques, baboons, and chimpanzees, both adult males and females display the "Westermarck effect": in sexual activity they spurn individuals with whom they were closely associated in early life. Mothers and sons almost never copulate, and brothers and sisters kept together mate much less frequently than do more distantly related individuals.

This elemental response was discovered, not in monkeys and apes but in human beings, by the Finnish anthropologist Edward A. Westermarck and first reported in his 1891 masterwork, *The History of Human Marriage.* The existence of the phenomenon has gained increasing support from many sources in the intervening years. None is more persuasive than the study of "minor marriages" in Taiwan by Arthur P. Wolf of Stanford University and his co-workers. Minor marriages, formerly widespread in southern China, are those in which unrelated infant girls are adopted by families, raised with the biological sons in an ordinary brother-sister relationship, and later married to the sons. The motivation for the practice appears to be to ensure partners for sons when an unbalanced sex ratio and economic prosperity combine to create a highly competitive marriage market among males for nubile females.

Across four decades, from 1957 to 1995, Wolf studied the histories of 14,200 Taiwanese women contracted for minor marriage during the late nineteenth and early twentieth centuries. The statistics were supplemented by personal interviews with many of these "little daughters-in-law," or *sim-pua,* as they are known in the Hokkien language, as well as with their friends and relatives.

What Wolf had hit upon was a controlled—if originally unintended—experiment in the psychological origins of a major piece of human social behavior. The *sim-pua* and their husbands were not biologically related, thus taking away all of the conceivable factors due to close genetic similarity. Yet they were raised in a proximity as intimate as that experienced by brothers and sisters in Taiwanese households.

The results unequivocally favor the Westermarck hypothesis. When the future wife was adopted before thirty months of age, she usually resisted later marriage with her de facto brother. The parents often had to coerce the couple to consummate the marriage, in some cases by threat of physical punishment. The marriages ended in divorce three times more often than "major marriages" in the same communities. They produced almost 40 percent fewer children, and a third of the women were reported to have committed adultery, as opposed to about 10 percent of wives in major marriages.

In a meticulous series of cross-analyses, Wolf and his co-workers identified the key inhibiting factor as close coexistence during the first thirty months of life of either or both of the partners. The longer and closer the association during this critical period, the stronger the later effect. The data allow the reduction or elimination of other imaginable factors that might have played a role, including the experience of adoption, financial status of the host family, health, age at marriage, sibling rivalry, and the natural aversion to incest that could have arisen from confusing the pair with true, genetic siblings.

A parallel unintended experiment has been performed in Israeli kibbutzim, where children are raised in crèches as closely as brothers and sisters in conventional families. The anthropologist Joseph Shepher and his co-workers reported in 1971 that among 2,769 marriages of young adults reared in this environment, none was between members of the same kibbutz peer group who had lived together since birth. There was not even a single known case of heterosexual activity, despite the fact that the kibbutz adults were not especially opposed to it.

From these examples, and a great deal of additional anecdotal evidence gleaned from other societies, it is evident that the human brain is programmed to follow a simple rule of thumb: *Have no sexual interest in those whom you knew intimately during the earliest years of your life.*

Is it possible that humans are not ruled by the Westermarck effect but instead simply use their intelligence and memory to recognize that sibling and parent-offspring incest create defective offspring? The answer is no. When the anthropologist William H. Durham examined the beliefs of sixty societies from around the world for references to any form of rational understanding of the consequences, he found only twenty with any degree of awareness. The Tlingit Amerindians of the Pacific Northwest, for example, grasped in a straightforward manner that defective children are often produced from matings of very close kin. Other societies not only knew that much but also developed folk theories to explain it. The Lapps of Scandinavia spoke of *mara*, the doom generated by partners in incest, as transmitted to their young. The Kapauku of New Guinea, in a similar perception, believed that the act of incest causes a deterioration of the vital substances. The people of Sulawesi, Indonesia, were more cosmic in their interpretation. They said that whenever people mate who have certain conflicting relationships, as between close kin, nature is thrown into confusion.

Curiously, while fifty-six of Durham's sixty societies had incest motifs in one or more of their myths, only five contained accounts of evil effects. A somewhat larger number ascribed beneficial results to transgressions, in particular the creation of giants and heroes. But even here incest was viewed as something special, if not abnormal.

The Westermarck effect is an epigenetic rule of gene-culture coevolution, in that it is the inherited predisposition of individuals to select and transmit through culture one out of multiple (in this case, two) options possible. Their parallel in medical genetics is the "susceptibility" genes of cancer, alcoholism, chronic depression, and many other of the more than a thousand known inherited diseases. Those who possess the genes are not absolutely condemned to acquire the trait, but in certain environments they are more likely than the average person to do so. If you are genetically prone to mesothelioma and you work in a building leaking asbestos dust, you are more likely than your co-workers to develop the disease. If you are genetically alcoholism-prone and socialize with heavy drink-

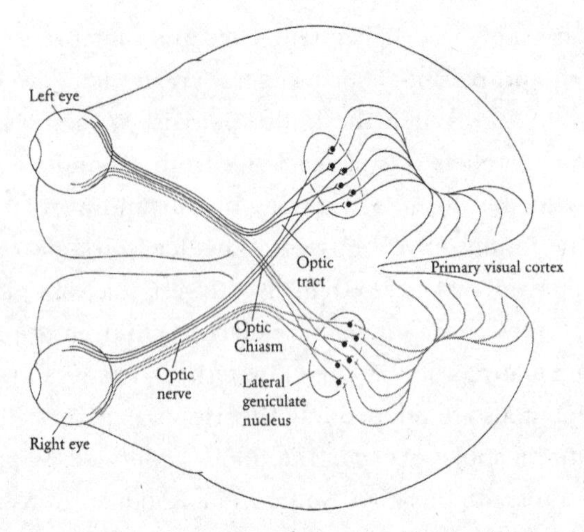

FIGURE 20-2. *The creation of color by the brain. Light frequencies are sorted in the retina into broad categories destined to be classified by the brain as colors. Neural impulses generated by the retina travel through the optic nerve to the lateral geniculate nuclei in the thalamus, a major transit and organizing center. From the thalamus, visual information travels to processing centers in the primary visual cortex and other brain regions. (Based on David H. Hubel and Torsten N. Wiesel, "Brain mechanisms of vision,"* Scientific American, *September 1979, p. 154.)*

ers, you are more likely than your genetically less-prone friends to become addicted. The epigenetic rules of behavior that affect culture, and have arisen by natural selection, act the same way but have the opposite effect. They are the norm, and strong deviations from them are likely to be scrubbed out by either cultural evolution or genetic evolution, or both. Seen in this light, both the genetic rules of gene-culture coevolution and disease susceptibility are consistent with the broad definition of "epigenetic" used by the U.S. National Institutes of Health as "changes in the regulation of gene activity and expression that are not dependent on gene sequence," including "both heritable changes in gene activity and expression (in the progeny of cells or individuals) and also stable, long-term alterations in the transcriptional potential of a cell that are not necessarily heritable."

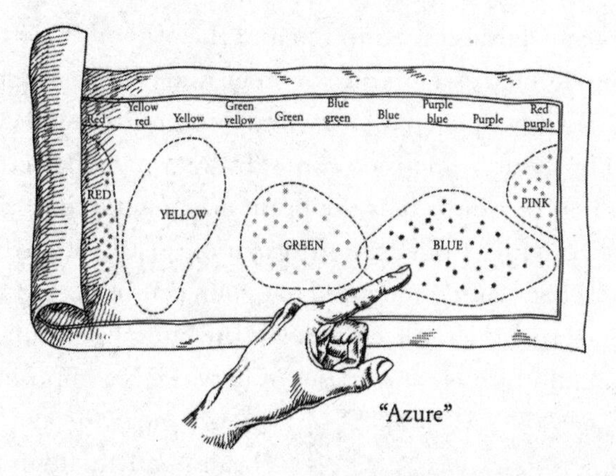

FIGURE 20-3. *The Berlin-Kay experiment, demonstrating that innate perception of primary colors guides the evolution of color vocabularies. Native language speakers concentrate their terms where color perception is most stable as light-wave frequency changes. (From Charles J. Lumsden and Edward O. Wilson,* Promethean Fire: Reflections on the Origin of Mind *[Cambridge, MA: Harvard University Press, 1983].)*

In a radically different category, a second case of gene-culture coevolution that has been equally well researched is color vocabulary. Scientists have traced it all the way from the genes that prescribe color perception to the final expression of color perception in language.

Color does not exist in nature. At least, it does not exist in nature in the form the untutored brain thinks. Visible light consists of continuously varying wavelengths, with no intrinsic color in it. Color vision is imposed on this variation by the photosensitive cone cells of the retina and the connecting nerve cells of the brain. It begins when light energy is absorbed by three different pigments in the cone cells, which biologists have labeled blue, green, or red cells according to the photosensitive pigments they contain. The molecular reaction triggered by the light energy is transduced into electrical signals that are relayed to the retinal ganglion cells forming the optic nerve. Here the wavelength information is recombined to yield signals distributed along two axes. The brain later interprets one axis as green to red and the other as blue to yellow, with yellow defined as a mixture of green and red. A particular ganglion cell, for example, may be excited by input from red cones and inhibited by input from green cones. How strong an electric signal is that it then transmits tells the brain how much red or green the retina is receiving. Collective information of this kind from vast numbers of cones and mediating ganglion cells is passed back into the brain, across the optic chiasma to the lateral geniculate nuclei of the thalamus, which are masses of nerve cells composing a relay station near the center of the brain, and finally into arrays of cells in the primary visual cortex at the extreme rear of the brain.

FIGURE 20-4. *In Paul Klee's* New Harmony *(1936), the eye is drawn first to the red squares, then tends to shift to other colors in a sequence roughly like the order followed in the evolution of color vocabularies. However, the possible connection between the physiological and cultural processes remains to be tested. (Paul Klee,* New Harmony [Neue Harmonie], *1936, oil on canvas, 36 7/8 x 26 1/8 inches [93.6 x 66.3 cm], Solomon R. Guggenheim Museum, New York, 71.1960.)*

Within milliseconds the visual information, now color-coded, spreads out to different parts of the brain. How the brain responds depends on the input of other kinds of information and the memories they summon. The patterns invoked by many such combinations, for example, may cause the person to think words denoting the patterns, such as "This is the American flag; its colors are red, white, and blue." Keep the following comparison in mind when pondering the seeming obviousness of human nature: an insect flying by would perceive different wavelengths, and break them into different colors or none at all, depending on its species, and if somehow it could speak, its words would be hard to translate into our own. Its flag would be very different from our flag, thanks to its insect (as opposed to human) nature. "This is the ant flag; its colors are ultraviolet and green" (ants can see ultraviolet, which we cannot see, but not red, which we can).

The chemistry of the three cone pigments—the amino acids of which they are composed and the shapes into which their chains are folded—is known. So is the structure of the DNA in the genes on the X chromosome that prescribe them, as well as that of the mutations in the genes that cause color blindness.

So, by inherited and reasonably well-understood molecular processes, the human sensory system and brain break the continuously varying wavelengths of visible light into the array of more or less discrete units we call the color spectrum. The array is arbitrary in an ultimately biological sense. It is only one of many arrays that might have evolved over thousands of millennia. But it is not arbitrary in a cultural sense. Having evolved genetically, it cannot be altered by either learning or fiat. All of the human culture traits involving color are derived from this unitary process. As a biological phenomenon, color perception exists in contrast to the perception of light intensity, the primary quality of visible light other than frequency. When we vary the intensity of light gradually, say by moving a dimmer switch smoothly up or down, we perceive the change as the continuous process it truly is. But if we use monochromatic light—project only one wavelength at a time—and change from one wavelength to the next in succession, we do not perceive such a continuity. What we see in going from the short-wavelength end to the long-wavelength end is first a broad band of blue (at least one band of wavelength more or less perceived as that color), then green, then yellow, and finally red. Add to the colors white, produced by the colors combined, and black, the absence of light.

The creation of color vocabularies worldwide is biased on this same biological constraint. In a famous experiment performed in the 1960s, Brent Berlin and Paul Kay tested the color concepts in native speakers of twenty languages, including Arabic, Bulgarian, Cantonese, Catalan, Hebrew, Ibibio, Thai, Tzeltal, and Urdu. The volunteers were asked to describe their color vocabulary in a direct and precise manner. They were shown a Munsell array, a spread of chips varying across the color spectrum from left to right, and increasing in brightness from the bottom to the top, and asked to place each of the principal color terms of their language on the chips closest to the meaning of the words. Even though the terms vary strikingly from one language to the next in origin and sound, the speakers placed them into clusters on the array that correspond, at least approximately, to the principal colors blue, green, yellow, and red.

The intensity of the learning bias was strikingly revealed by an experiment conducted on color perception during the late 1960s by Eleanor Rosch. In looking for "natural categories" of cognition, Rosch exploited the fact that the Dani people of New Guinea have no words to denote color; they speak only of *mili* (roughly, "dark") and *mola* ("light"). Rosch considered the following question: If Dani adults set out to learn a color vocabulary, would they do so more readily if the color terms correspond to the principal innate hues? In other words, would cultural innovation be channeled to some extent by the genetic constraints? Rosch divided sixty-eight volunteer Dani men into two groups. She taught one a series of newly invented color terms placed on the principal hue categories of the array (blue, green, yellow, red), where most of the natural vocabularies of other cultures are located. She taught a second group of Dani men a series of new terms placed off center, away from the main clusters formed by other languages. The first group of volunteers, following the "natural" propensities of color perception, learned about twice as quickly as those given the competing, less natural color terms. They also selected these terms more readily when allowed a choice.

Now comes the question that must be answered to complete the transit from genes to culture. Given the genetic basis of color vision and its general effect on color vocabulary, how great has been the dispersion of traits among different cultures? We have at least a partial answer. In the case of the Westermarck effect and the incest avoidance it creates, all societies are almost completely consistent. However, color vocabularies are very different in this regard. A few societies are relatively unconcerned with color, getting along with a rudimentary classification. Others make many fine distinctions in hue and intensity within each of the basic colors. They have spaced their vocabularies out.

Has the spacing out of color terms been random? Evidently not. In later investigations, Berlin and Kay observed that each society uses from two to eleven basic color terms, which are focal points spread across the four elementary color blocks perceived in the Munsell array. The full complement, to use the English-language terminology, is black, white, red, yellow, green, blue, brown, purple, pink, orange, and gray. Each can be equated across cultures with one color term out of the eleven or some combination of terms. When we say "pink," for example, there may be in another given language, an equivalent term or, say, a term that means to us "pink" and/or "orange." The Dani language, for instance, uses only two of the terms, the English language all eleven. In passing from societies with simple classifications to those with complicated classifications, the combinations of basic color terms as a rule grow in the following hierarchical fashion:

> Languages with only two basic color terms use them to distinguish black and white.
>
> Languages with only three terms have words for black, white, and red.
>
> Languages with only four terms have words for black, white, red, and either green or yellow.
>
> Languages with only five terms have words for black, white, red, green, and yellow.

Languages with only six terms have words for black, white, red, green, yellow, and blue.

Languages with only seven terms have words for black, white, red, green, yellow, blue, and brown.

No such precedence occurs among the remaining four basic colors, purple, pink, orange, and gray, when these have been added on top of the first seven.

If basic color terms were combined at random, which is clearly not the case, human color vocabularies would be drawn helter-skelter from among a mathematically possible 2,036 sequences. The Berlin-Kay progression suggests that for the most part they are drawn from only twenty-two.

Subsequent new work has confirmed the reality of the eleven basic words for color, such that those of one language can be matched with those of other languages—whether one each in turn to one, many to one, or one to many. However, precisely where the terms are placed in each of the focal colors differs among languages. The positioning appears to depend on the importance of the color at the point of the basic focal area where it is placed. It also depends on how well the placement distinguishes the basal color from the one next to it.

A fundamental question concerning gene-culture coevolution evolved by the relation between color categories and language is the extent to which one affects the other. An influential hypothesis effectively expressed by Benjamin Lee Whorf in the late 1930s and early 1940s suggests that language not only serves to communicate what we perceive in the rest of the world but also influences what we literally perceive. In the case of color vocabularies, the body of research to date has come to favor a middle view, that the brain does filter and distort true color in some ways but does not exclusively determine its categories.

Direct evidence concerning the relation of color to language has been recently obtained from MRI studies of brain activity. The perception of color categories is more strongly correlated with the right visual field of the brain. When subjects were shown various sequences of color categories, their pattern of brain activity was stronger in the right visual field for colors in different color categories than for the same color category—as expected. But different color categories also provoked stronger activation in the left hemisphere language region. This result suggests that the language regions provides some amount of top-down control of activity in the visual cortex.

Evolutionary biologists on their part have begun to probe the question of why human cultures in general select a particular sequence of color categories as they add terms to their repertory. One promising candidate for surmise is the dominance of the color red, which makes its appearance early in the evolutionary sequence. A likely explanation, according to André A. Fernandez and Molly R. Morris, is that red and orange are colors characteristically found in fruit. Early arboreal primates would find advantage in moving toward this color in the midst of an almost entirely brown and green environment. As some species became social, the hypothesis continues, they chose these colors to advertise their sexual readiness. In the general theory of instinct evolution, red and reddish hues were "ritualized" in ancestral Old World primates to serve in visual communication.

· 21 ·

How Culture Evolved

IN THE GOUALOUGO TRIANGLE forest of the Congo, a chimpanzee breaks a thin branch from an understory sapling, pulls off its leaves, and pokes it into a nearby termite mound. Inside the mound, the soft white workers flee from the branch, while soldier termites rush forward to seize hold of the branch with their needle-pointed mandibles. They hold on to the stick in a grip of death. The chimpanzee knows this. He waits briefly until a mass of defenders has accumulated, then pulls the stick up, strips off the soldiers, and eats them. This practice does not occur everywhere. It is part of local chimpanzee culture in some populations but not others, learned by one individual watching another.

In the land of the Yanomamo, between the Rio Negro and the Rio Branco in a region that overlaps Brazil and Venezuela, a small group of villagers leave a collective house and walk to a stream three kilometers away. They drop timbó poison into the water, wait, and collect the fish that float to the surface. The catches are carried home to be shared with others in the village. This practice occurs in the summer season. At other times women come singly to the stream. They catch fish with their hands and bite them in the neck to kill them. Off the coast of Alaska, at a very different level, professional deep-sea fishermen drop long lines bearing rows of hooks onto the floor of the Pacific Ocean, at depths of 3,600 feet or more. They bring up sablefish, also known as black cod or butterfish (or gindara when turned into sushi). The catch is cleaned and refrigerated, transported to markets on the coast, and distributed worldwide to high-end restaurants and private tables.

The practice of fishing is a particular culture that has evolved over what has likely been millions of years, extremely slow at the beginning and then faster and still faster and finally explosively fast. The route to a dinner of butterfish is only one of myriad cultural categories that have streamed forth from the mind of man, branched, and anastomosed since the dawn of the Neolithic era, finally coming together to create the substance of modern global civilization. We did not invent culture. The common ancestors of chimpanzees and prehumans invented it. We elaborated what our forebears evolved to become what we are today.

As defined broadly by both anthropologists and biologists, culture is the combination of traits that distinguishes one group from another. A culture trait is a behavior that is either first invented within a group or else learned from another group, then transmitted among members of the group. Most researchers also agree that the concept of culture should be applied to animals and humans alike, in order to stress its continuity from one to the other and notwithstanding the immensely greater complexity of human behavior.

The most advanced cultures known to occur in animals are those of the chimpanzees and their close relatives, the bonobos. Comparative studies of chimpanzee populations scattered across Africa have revealed a surprising number of culture traits, and differences in the combinations of such traits found from one population to the next.

The role of imitation of one group member by another in the spread of culture traits has been supported by experiments with two chimp colonies. In the procedure, researchers selected a high-ranking female from each of the two groups and gave her a private demonstration on how to obtain food from a specially designed container. With food as the reward, the chimpanzees proved quick studies. One learned a "poke" technique and the other a "lift" technique. When returned to their own groups, each continued to practice the method shown her. A large majority of her companions soon began using the same method of container opening. The spread may have been a direct imitation of the teacher chimp, but it is equally possible that the students learned instead by watching the mechanical motions of the food dispenser. If the latter proves true, further studies may reveal social learning to be very different in chimps from that in humans.

The occurrence of authentic culture has also been convincingly documented in orangutans and dolphins. A striking example of innovation and cultural transmission in the latter animals is sponge-fishing by the bottlenose dolphins of Australia's Shark Bay. A small minority of females attach a fragment of sponge to their nose, then push with it to flush fish from tight hiding places on the bottom of the bay channels. Culture in dolphins should not come as a great surprise. They are among the most intelligent of all animals, ranking in that respect just below monkeys and apes. Because dolphins are also intensely imitative during their social interactions, it seems very likely that the Shark Bay innovators engage in true cultural transmission. Then why haven't dolphins and other brainy cetaceans, whose evolution extends back millions of years, progressed further in social evolution? Three reasons stand out. Unlike primates, they have no nests or campsites. They have flippers for forelimbs. And in their watery realm, controlled fire is forever denied.

The elaboration of culture depends upon long-term memory, and in this capacity humans rank far above all animals. The vast quantity stored in our immensely enlarged forebrains makes us consummate storytellers. We summon dreams and recollections of experience from across a lifetime and use them to create scenarios, past and future. We live in our conscious mind with the consequence of our actions, whether real or imagined. Placed out in alternative versions, our inner stories allow us to override immediate desires in favor of delayed pleasure. By long-range planning we defeat, for a while at least, the urging of our emotions. This inner life is why each person is unique and precious. When one dies, an entire library of both experience and imaginings is extinguished.

How much does death extinguish? I believe I am typical in conceiving how much. On occasion I close my eyes and return in remembrance to Mobile and the nearby Alabama Gulf Coast as they were in the 1940s. Arriving there, a boy once more, I travel from one end of the surrounding county to the other, on my single-gear, balloon-tire Schwinn bicycle. More detail follows vividly. I remember my extended family, each one in a network of people of his or her own, each with memories shared in part with others. They existed in what must have seemed to them to be the center of the world at the center of time. They lived as though Mobile as it was then would never change by much. Everything mattered, every detail, at least for a while. Somehow, in one form or another everything collectively remembered was important to someone. Now these people are all gone. Almost everything held in their vast collective memory is forgotten. I know that when I die my memories and with them this earlier world, and the immensity of knowledge it contained, will also be gone. But I know further that all those networks, and all that library of remembrance, even though vanished, were vital to a part of humanity. They are why I survived, and went on.

Animals also have long-term memories, which serve them well in survival. Pigeons can manage the memorization of up to 1,200 pictures. Clark's nutcrackers, a bird species that in nature stores acorns in the manner of squirrels, remembered when tested in captivity up to 25 caches in a room containing 69 caches, and held the memories for as long as 285 days. Both of these bird species, not surprisingly, are exceeded by baboons. Tests have revealed that these obviously intelligent primates can memorize at least 5,000 items and retain them for at least three years. Human long-term memory is, in its turn, vastly greater than that of any animal known. No method to my knowledge has been devised to measure the capacity in an individual human being, even to the nearest order of magnitude.

The great gift of the conscious human brain is the capacity—and with it the irresistible inborn drive—to build scenarios. For each story in turn, the conscious mind summons only a minute fraction of the brain's accumulated long-term memory. How this is done remains controversial. One group of neuroscientists argues that fragments of long-term memory are transformed from long-term storage and congealed into working memory to make scenarios. A second school believes, with the same data, that the process is achieved simply by the arousal of long-term memory—with no transfer from one sector of the brain to another needed.

Either way, it is clear that during a relatively swift three million years of evolution the genus *Homo* generated something never before approached by any other kind of animal: a memory bank held in a huge brain cortex of over ten billion neurons, each neuron extending an average of 10,000 branches that connect with other such cells. These linkages, the basic units of brain tissue, form intricate pathways of circuits and integrating relay stations. Networks of pathways and relay stations, sometimes called modules, somehow organize all of the instincts and memory of a human brain.

At first, the immense complexity in brain architecture created a difficult problem for theoretical models of genetics applied to evolutionary theory. The human genome contains as few as 20,000 protein-coding genes. Of these, only a fraction prescribe our sensory and nervous systems. The problem posed is this: How could cellular architecture so complicated be programmed with so few genes?

The gene-shortage dilemma has been solved by a concept originating in developmental genetics. Multiple modules, researchers have found, can be built by instructions that first replicate them from a single program, followed by separate programs (and separate genes) that command each module tissue to specialize according to its location in the brain. Further specialization can be achieved by the input received from the environment outside the brain. In a simple parallel, a centipede does not need an ensemble of hundreds of genes to program the development of its hundred pairs of legs. Only several will do. A great deal remains to be learned about the genetic control of brain development, but at least the theoretical capability of human genes to accomplish it has been demonstrated.

With the genetic coding for the development of the human brain no longer an overwhelming puzzle, we can turn to the origin of mind and language. Scientists long ago abandoned the idea of the brain as a blank slate upon which all of culture is inscribed by learning. In this archaic view, all that evolution has achieved is an exceptional ability to learn, based upon an extremely large capacity for long-term memory. A different view now prevails: the brain has a complex inherited architecture. As a consequence of the way it was built, the conscious mind, one of the architecture's products, originated by gene-culture coevolution, an intricate interplay between genetic and cultural evolution.

Archaeologists have joined geneticists and neuroscientists in the effort to understand the evolutionary origin of language and mind. In order to retrace the steps and timing of these elusive events, they have initiated a new field of study called "cognitive archaeology." At first, such a hybrid discipline might seem to have little chance for success. After all, other than exhumed bones the only evidence left by ancient humans consists of the ash of campfires, fragments of tools, discarded remnants of meals, and other refuse. Nonetheless, by new methods of analysis and experimentation, researchers have been able to conclude this much: abstract thought and syntactical language emerged no later than 70,000 years ago. The key to this conclusion lies in the existence of certain artifacts, and in deductions of the mental process required to manufacture the artifacts. Of special importance in the mode of reasoning is the hafting of stone points onto the ends of spears. The practice was begun as long as 200,000 years ago by both the Neanderthal people of Europe and early *Homo sapiens* of Africa. This in itself was a significant technological invention, yet still it tells us little about reasoning and communication. By 70,000 years ago, however, a major new advance had been achieved by *Homo sapiens* which, when recently analyzed, shed light on cognitive evolution. Hafting, the study concluded, had become far more sophisticated. A series of steps was used to build spears, from firing and shaping the knapped stone tip to the use of

FIGURE 21-1. *That Neanderthal culture did not advance significantly during the history of the species is likely due to the inability to link domains of intelligence to create new abstract patterns and to imagine complex scenarios. (From Steven Mithen, "Did farming arise from a misapplication of social intelligence?"* Philosophical Transactions of the Royal Society ***B*** *362: 705–718 [2007].)*

acacia gum, beeswax, and other artifacts to hold the tip in place. What this tells us about cognition has been nicely summarized by Thomas Wynn:

> The artisans needed to understand the properties of their ingredients (e.g., cohesiveness), to be able to judge the effects of temperature, to be able to switch attention back and forth between separate rapidly changing variables, and to be flexible enough to adjust to the variability inherent in naturally occurring ingredients.

And what of speech? A conscious mind able to generate abstractions and piece them together in a complex scenario might, it seems, also generate a syntactical language, with sequences of subject, verb, and object.

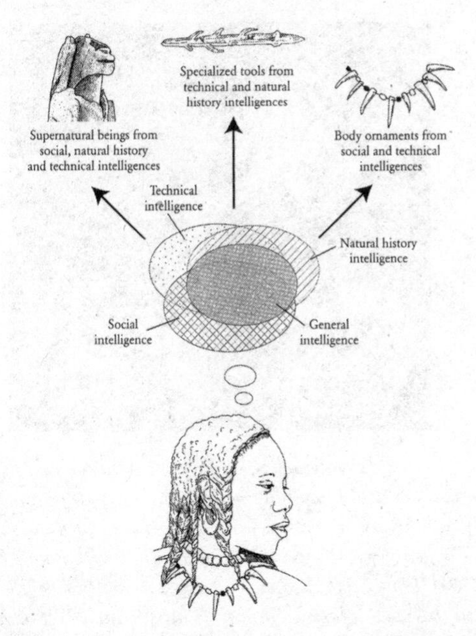

FIGURE 21-2. *The advance of Late Paleolithic intelligence and culture of* Homo sapiens *is suggested here. The remarkable advance of Late Paleolithic human culture was evidently due to the ability to link stored memory in different domains to create new forms of abstraction and metaphor. (From Steven Mithen, "Did farming arise from a misapplication of social intelligence?"* Philosophical Transactions of the Royal Society ***B** 362: 705–718 [2007].)*

In searching for the ancient origins of any species, it is customary to turn to comparative biology in order to learn how other, closely related species lived and might have evolved. The search for the genesis of the human mind has brought scientists for a close look at the Neanderthals (*Homo neanderthalensis*), about whom we have come to know a great deal. Modern humanity's sister species occupied Europe throughout the time *Homo sapiens* was achieving its advanced cognitive powers in Africa. It persisted there for more than 200,000 years. The last Neanderthal of which we have record died approximately 30,000 years ago in southern Spain. The species had almost certainly been pushed to extinction by *Homo sapiens*

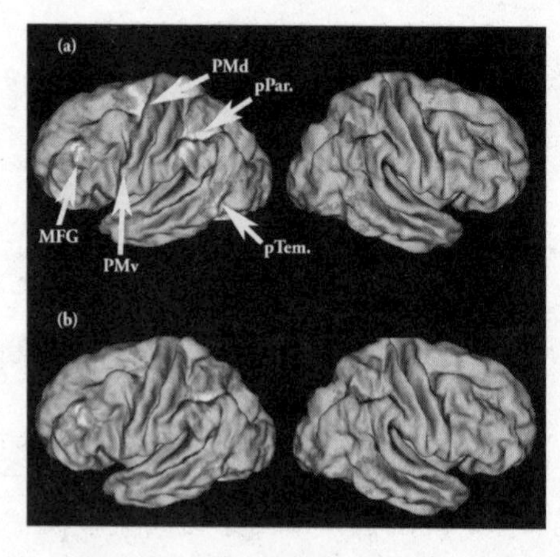

FIGURE 21-3. *The complex interaction of different mental domains in the modern human brain is illustrated by activity in different parts of the brain while an adult (a) thought about the case of tool use and (b) communicated the same tool with pantomime. The activity maps were made with functional magnetic resonance imaging (fMRI). (From Scott H. Frey, "Tool use, communicative gesture and cerebral asymmetries in the modern human brain,"* Philosophical Transactions of the Royal Society ***B*** *363: 1951–1957 [2008].)*

when that more adaptable species spread gradually north and west across the European continent.

At first it was a fair contest. The Neanderthals started neck and neck with their *sapiens* counterparts while the latter were still in Africa. Their stone tools were at first as sophisticated as those of *sapiens.* Their knives had straight sharp edges, probably used for scraping. Others had serrated edges likely used for sawing. Sharp-pointed pieces were hafted in a simple manner to staffs to make spears. The Neanderthal toolkit appears designed for the life the species led as big-game hunters. Neanderthals evidently moved around a great deal, as expected of carnivore specialists. They cooked and perhaps also smoked meat, wore clothing, and kept warm at their meager campsites in the bitter cold of winter with the help of fire. From the recent sequencing of their genetic code, an extraordinary scientific achievement in its own right, we know they possessed the FOX2 gene,

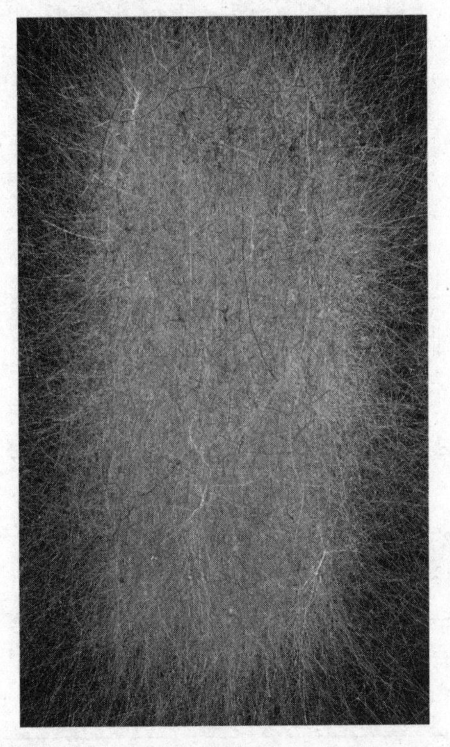

FIGURE 21-4. *The immense complexity of the human brain can be imagined by this model of the 100,000 neurons, in a slice half a millimeter by two millimeters in size, from a two-week-old rodent brain. Basic computational units of this kind are repeated millions of times in the human brain. (Jonah Lehrer, "Blue brain,"* Seed, *no. 14, pp. 72–77 [2008]. From research by Henry Markham et al., École Polytechnique Fédérale de Lausanne.)*

associated with language ability, and in a particular code sequence uniquely shared with *Homo sapiens*. Thus they may well have had a language. At maturity the Neanderthal brains were slightly larger on average than those of *Homo sapiens*. The brains of their infants and children also grew faster than those of *sapiens*.

Neanderthals are fascinating in every respect as another human species parallel to *Homo sapiens*—an evolutionary experiment available for comparison with our own. Yet perhaps the most interesting thing about them is not what they were but what they failed to

TABLE 21-1. The cultures of different wild chimpanzee groups in Africa are defined by their combinations of socially learned behaviors. [Based on the summary by Mary Roach, "Almost Human," *National Geographic* (April 2008), pp. 136–137.]

	Use rocks to smash open nuts and fruit	*Chew leaf to make sponge*	*Go into water*	*Fish termite with twigs*	*Rain dance*	*Throw rocks*	*Hunt small animals*	*Groom others with hands clasped over heads*	*Clip leaves with teeth for attention*
Assirik, Senegal	x	x	-	-	?	x	x	x	-
Fongoli, Senegal	x	x	x	x	x	x	x	x	x
Guinea Bossou	x	x	x	-	x	x	x	-	x
Taï Nat. Pk., Côte d'Ivoire	x	x	-	-	x	x	x	x	x
Goualougo Triangle, Congo	-	x	-	x	x	-	-	x	-
Budongo, Uganda	-	x	-	-	x	-	x	-	-
Kibale Nat. Pk., Uganda	-	x	-	-	x	x	x	x	x
Gombe Nat. Pk., Tanzania	x	x	-	x	x	x	x	x	-
Mahale-K, Tanzania	-	-	-	x	x	x	x	x	x
Mahale-M, Tanzania	-	x	x	x	x	x	x	x	x

FIGURE 21-5. *The mammoth steppe, theater of culture's creative explosion, is preserved in valley grasslands and mountain forests similar to these in the present-day Arctic National Wildlife Refuge. During the ice age, early* Homo sapiens *advanced across Eurasia south of the continental glacier, hunting large animals and replacing its sister species* Homo neanderthalensis. *("The Oneiric Autumn," from* Arctic Sanctuary: Images of the Arctic National Wildlife Refuge *[Fairbanks: University of Alaska Press, 2010], p. 115. Photographs by Jeff Jones, essays by Laurie Hoyle.)*

become. Virtually no progress occurred in their technology or culture during their two hundred millennia of existence. No tinkering with tool manufacture, no art, and no personal decoration—at least none exists in the archaeological evidence we have so far.

Homo sapiens meanwhile pressed forward, and at about the time Neanderthals left the scene the cognitive achievements of *sapiens* flowered dramatically. The first population worked north along the Danube into the European heartland about 40,000 years ago. Ten thousand years later, the innovations marking the Late Paleolithic era had begun: elegant representational cave art; sculpture, including a lion's head on a human body; bone flutes; controlled burning with corrals to direct and capture game; and costumed shamans.

What catapulted *Homo sapiens* to this level? Experts on the subject agree that increased long-term memory, especially that put into working memory, and with it an ability to construct scenarios and plan strategy in brief periods of time, played the key role in Europe and elsewhere, both before the African breakout and afterward. What was the driving force that led to the threshold of complex culture? It appears to have been group selection. A group with members who could read intentions and cooperate among themselves while predicting the actions of competing groups, would have an enormous advantage over others less gifted. There was undoubtedly competition among group members, leading to natural selection of traits that gave advantage of one individual over another. But more important for a species entering new environments and competing with powerful rivals were unity and cooperation within the group. Morality, conformity, religious fervor, and fighting ability combined with imagination and memory to produce the winner.

· 22 ·

The Origins of Language

THE EXPLOSION OF INNOVATIONS that lifted humanity to world dominance surely did not result from a single empowering mutation. Even less likely did it come as some mystic afflatus that descended upon our struggling forebears. Nor could it have been due to the stimulus of new lands and rich resources—enjoyed also by the relatively unprogressive species of horses, lions, and apes. Most probably it was the gradual approach to and final attainment of a tipping point, the crossing over of a threshold level of cognitive ability that endowed *Homo sapiens* with a dramatically high capacity for culture.

The climb had begun in Africa at least two million years earlier, with the habiline precursors of *Homo erectus.* At that point the forebrain began its phenomenal growth, not seen in any other complex structure during half a billion previous years of animal evolution. What ignited this change? The preadaptations for eusociality, the most advanced level of social organization, had all been laid in place, but such was also true for the multiple species of australopithecines that existed up to that time, none of whom hit upon the path to rapid cerebral growth. The clue to the advance to *Homo,* I believe, lies in the critical preadaptation that had carried the few other evolving animal species in the history of life that have managed to cross the eusociality threshold. Every one, without exception, from the two dozen or so insect and crustacean lines to the naked mole rats, defended a nest from which members could forage for enough food to sustain the colony. In the rare instances where such colonies could outcompete solitary individuals, they remained at the nest instead of dispersing to renew the cycle of solitary life.

It is no coincidence that by the origin of *Homo erectus*, and very likely earlier, at the time of its immediate ancestor *Homo habilis*, small groups had begun to establish campsites. They were able to create these equivalents of animal nests because they had shifted their diet from vegetarian to omnivore, with a substantial reliance on meat. They scavenged and hunted, and in time they came to rely on the high caloric yield of cooked animal flesh. The archaeological evidence indicates that no longer did their bands wander constantly through a territory gathering fruit and other vegetable food, in the manner of contemporary chimpanzees and gorillas. Now they selected defensible sites and fortified them, with some staying for extended periods to protect the young while others hunted. When controlled fire at the camp was added, the advantage of this way of life was solidified.

Still, meat and campfire are not enough by themselves to explain the rapid increase in size of the brain that occurred. For the missing piece we can turn, I believe with some confidence, to the cultural intelligence hypothesis of Michael Tomasello and his co-workers in biological anthropology, developed during the past three decades.

These researchers point out that the primary and crucial difference between human cognition and that of other animal species, including our closest genetic relatives, the chimpanzees, is the ability to collaborate for the purpose of achieving shared goals and intentions. The human specialty is intentionality, fashioned from an extremely large working memory. We have become the experts at mind reading, and the world champions at inventing culture. We not only interact intensely with one another, as do other animals with advanced social organizations, but to a unique degree we have added the urge to collaborate. We express our intentions as appropriate to the moment and read those of others brilliantly, cooperating closely and competently to build tools and shelters, to train the young, to plan foraging expeditions, to play on teams, to accomplish almost all we need to do to survive as human beings. Hunter-gatherers and Wall

Street executives alike gossip at every social gathering, evaluating others, estimating their truthfulness, and predicting their intentions. Our leaders spin political strategy with the crafts of social intelligence. Businessmen strike deals from intention reading, and the bulk of the creative arts is devoted to its expression. As individuals we can live scarcely a day without the exercise of cultural intelligence, even if only in the frequent rehearsals that invade our private thoughts.

Human beings are enmeshed in social networks. Like the proverbial fish in the sea, we find it difficult to conceive of any place different from this mental environment we have evolved. From infancy we are predisposed to read the intention of others, and quick to cooperate if there is even a trace of shared interest. In one revealing experiment, children were shown how to open the door to a container. When adults tried to open the door but pretended not to know how, the children stopped what they were doing and crossed the room to help. Chimpanzees put in the same circumstance, but far less advanced in cooperative awareness, made no such effort.

In another experiment, the chimpanzees were given tests of intelligence, and their scores compared with those of 2.5-year-old children tested before schooling and literacy. In solving physical and spatial problems (for example, locating a hidden reward, discriminating different quantities, understanding the properties of tools, using a stick to reach an object out of reach), the chimpanzees and young children were about equal. On the other hand, the children displayed more advanced skills than the chimpanzees in a variety of social tests. They learned more while watching a demonstration, better understood cues that aid in locating a reward, followed the gaze of others to a target, and grasped the intention of the actions of others in searching for a reward. Humans, it appears, are successful not because of an elevated general intelligence that addresses all challenges but because they are born to be specialists in social skills. By cooperating through the communication and the reading of intention, groups accomplish far more than the effort of any one solitary person.

The early populations of *Homo sapiens*, or their immediate ancestors in Africa, approached the highest level of social intelligence when they acquired a combination of three particular attributes. They developed shared attention—in other words, the tendency to pay attention to the same object at ongoing events as others. They acquired a high level of the awareness they needed to act together in achieving a common goal (or thwarting others in the attempt). And they acquired a "theory of mind," the recognition that their own mental states would be shared by others.

When these qualities had been sufficiently developed, languages comparable to those that prevail today were invented. This advance certainly occurred before the African breakout 60,000 years ago. By that time, the colonists had the full linguistic capability of their modern descendants and probably used sophisticated languages. The chief evidence for this conclusion is that present-day aboriginal populations, direct descendants of the colonists now existing in settled relict populations from Africa to Australia, all possess languages of such high quality and the mental attributes necessary to invent them.

Language was the grail of human social evolution, achieved. Once installed, it bestowed almost magical powers on the human species. Language uses arbitrarily symbols and words to convey meaning and generate a potentially infinite number of messages. It is capable ultimately of expressing to at least a crude degree everything the human senses can perceive, every dream and experience the human mind can imagine, and every mathematical statement our analyses can construct. It seems logical that language did not create the mind, but the opposite. The sequence in cognitive evaluation was from intense social interaction in early settlements to a synergism with increasing ability to read and act upon intention, to a capacity to create abstraction in dealing with others and the outside world and, finally, to language. The rudiments of human language might have appeared as the essential enabling mental qualities that came together and coevolved in a synergistic fashion. But it is highly unlikely that it preceded them. Michael Tomasello and his coauthors have stated the case as follows:

> Language is not basic; it is derived. It rests on the same underlying cognitive and social skills that lead infants to point to things and show things to other people declaratively and informatively, in a way that other primates do not do, and that lead them to engage in collaborative and joint attentional activities with others of a kind that are also unique among primates. The general question is, What is language if not a set of coordination devices for directing the attention of others? What could it mean to say that language is responsible for understanding and sharing intentions, when in fact the idea of linguistic communication without these underlying skills is incoherent. And so, while it is true that language represents a major difference between humans and other primates, we believe that it actually derives from the uniquely human abilities to read and share intentions with other people—which also underwrite other uniquely human skills that emerge along with language such as declarative gestures, collaboration, pretense, and imitative learning.

Animals are occasionally described as having a language. Honeybees, perhaps the most striking example, are said to communicate with abstract signals during their dances on the combs of the hive, as well as on the massed bodies of their fellow workers during emigration to new nest sites. The dancing bee does indeed convey the direction and distance of the target, whether a source of nectar and pollen or a potential new nest site. But the code is fixed, and has been for probably millions of years. Also, the dance is not an abstract symbol as composed in human words and sentences. It is a reenactment of the flight the outbound bees must take to get to the target. If the dancer moves in a circle, it means the target is close to the nest ("travel closely around the nest to find the target"). The waggle dance, tracing a figure eight repeated over and over, tells of a more distant target. The middle segment of the 8, more like the Greek letter Θ, is the direction to take with reference to the angle of the sun, and the length of the middle segment is proportional to the distance to the target. This is impressive, but only humans can say something like, "Go out the entrance, turn right, keep on the road until you get to the first light, then look for the restaurant halfway down the block—no wait, it's on the next corner."

Unlike communication in bees and other animals, human language became capable of detached representation, in which reference is made to objects and events not present in the immediate vicinity—or even in existence. Further, human speech adds information by prosody, the emphasis on particular words and the pacing of their flow in order to invoke mood, to highlight emphasis, or denote one meaning of a phrase as opposed to another. Human language is shot through with irony, a fine-tuned play of hyperbole and misdirection that conveys a meaning different from that in the phrase as literally worded. Language can be indirect, insinuating a message instead of stating it baldly, and thereby leaving open plausible deniability. Examples include overt, even clichéd sexual come-ons ("Would you like to come up and see my etchings?"); polite requests ("If you could help me change this flat tire, I'd be eternally grateful"); threats ("Nice store you got here. Be a shame if something happened to it"); bribes ("Gee, officer, would it be possible for me to pay the ticket right here?"); soliciting for a donation ("We hope you will join our Leadership Program"). As explained by Steven Pinker and other scholars of the subject, indirect speech has two functions, to convey information and to negotiate a relationship between the speaker and the hearer.

Because language is central to human existence, it is important to know its evolutionary history. In pursuing that goal, we are hampered by the fact that language is also the most perishable of artifacts. Archaeological evidence goes back only to the origin of writing, about five thousand years ago, by which time the critical genetic changes in *Homo sapiens* had occurred and the sophisticated rules of speech were uniformly in place in all societies worldwide.

Even so, there exist a few patterns in speech that can be cited as products of evolution. One such vestige is turn-taking during conversations. A long-standing popular impression is that cultures differ in the length of the gap between turns. Nordics, for example, are thought to take long pauses between one person's speaking and the other's answering. New York Jews, as comedians have depicted them, are thought to have a preference for nearly simultaneous speech. However, when the conversational gaps of speakers of ten languages from around the world were actually measured, all were shown to avoid overlap (but not interruption), and the length of the turnover gaps was found to be almost the same. On the other hand, conversations between speakers of different languages yielded considerable variation in the gap, as conversationalists struggled to grasp meaning and intention. This understandable effect is probably the source of the perception that cultures differ in the pace of conversation.

Another vestige of early linguistic evolution recently documented is in nonverbal vocalizations, the utterances of which are probably older than language. Vocalizations that communicate negative emotions (anger, disgust, fear, and sadness) were found, for example, to be the same between native speakers of English in Europe and speakers of the Himba language, the latter limited to remote and culturally isolated settlements in northern Namibia. In contrast, nonverbal vocalizations that communicate positive emotions (achievement, amusement, sensual pleasure, and relief) do not match in the same way. The reason for the difference is unknown.

The fundamental question concerning the origin of language is not conversational turn-taking and prelingual utterances, however, but grammar. Is the order in which words and phrases are strung together learned, or in some manner innate? In 1959, a historic exchange occurred between B. F. Skinner and Noam Chomsky on this subject. It took the form of a long essay review by Chomsky of Skinner's book *Verbal Behavior*, published in 1957. Skinner, the founder of behaviorism, said language is all learned. Chomsky disagreed. Learning a language, he said, with all its grammatical rules added, is too complex for a child to memorize during the time available. Chomsky at first appeared to win the argument. He subsequently reinforced his point by bringing forth a series of rules that, he proposed, are followed spontaneously in the developing brain.

These rules were, however, expressed in an almost incomprehensible manner, an unfortunate example of which follows:

> To summarize, we have been led to the following conclusions, on the assumption that the trace of a zero-level category must be properly governed.
>
> 1. VP is α-marked by I.
> 2. Only lexical categories are L-markers, so that VP is not L-marked by I.
> 3. α-government is restricted to sisterhood without the qualification (35).
> 4. Only the terminus of an X^0-chain can α-mark or Case-mark.
> 5. Head-to-head movement forms an A-chain.
> 6. SPEC-head agreement and chains involve the same indexing.
> 7. Chain coindexing holds of the links of an extended chain.
> 8. There is no accidental coindexing of I.
> 9. I-V coindexing is a form of head-head agreement; if it is restricted to aspectual verbs, then base-generated structures of the form (174) count as adjunction structures.
> 10. Possibly, a verb does not properly govern its α-marked complement.

Scholars struggled to understand what appeared to be a profound new insight into the workings of the brain (I was one of them, in the 1970s). Deep grammar or universal grammar, as it was variously called, was a favorite topic of befuddled *salonistes* and college seminars. For a long time, Chomsky succeeded because, if for no other reason, he seldom suffered the indignity of being understood.

Eventually, analysts were able to put into comprehensible language and diagrams what Chomsky and his followers were saying. Among the most accessible and sympathetic was Steven Pinker's best-selling *The Language Instinct* (1994).

Yet, even with Chomsky decoded, the question remained: Is there really a universal grammar? An overwhelmingly powerful instinct to learn language certainly exists. There is also a sensitive period in a child's cognitive development when the learning is quickest. In fact, so swift is language acquisition, so fierce the child's effort to learn, that Skinner's argument may not be so dismissible after all. Perhaps there is a time in early childhood, and the ability to learn words and word order so efficient, that a special brain module for grammar is not a necessity.

In fact, as experimental and field research has progressed in recent years, a view of the evolution of language different from "deep grammar" has emerged. The alternative allows for epigenetic rules, entailing "prepared learning," in the way languages of individual cultures evolve. But the constraints imposed by these rules are very broad. The psychologist and philosopher Daniel Nettle has described the emergence and the possibilities it offers for new directions in research on linguistics:

> All human languages perform the same function, and the set of distinctions they use to do so is probably highly constrained. The constraints come from the universal architecture of the human mind, which influences language form through the way it hears, articulates, remembers, and learns. However, within these constraints, there is latitude for variation from language to language. For example, the major categories of subject, verb, and object vary in their typical order, and some languages signal grammatical distinctions primarily by syntax, or the combinatorics of words, whereas others achieve this mainly through morphology, or the internal mutation of words.

There now exist a number of likely new avenues for penetrating more deeply into the language enigma, pulling linguistics away from the contemplation of sterile diagrams and more in the direction of biology. One is the manner in which the external environment opens or narrows the constraints in language evolution, whether by genetic evolution or cultural evolution, or both. In warm climates, to take a simple example, languages around the world have evolved to use more vowels and fewer consonants, creating more sonorous combinations of sounds. The explanation for the trend may be a simple matter of acoustic efficiency. Sonorous sounds carry further, in accord with the tendency of people in warm climates to spend more time outdoors and keep greater distances apart.

Another factor in the generation of language diversity may be genetic. There is a correlation in geographical patterns between the use of voice pitch to convey grammar and word meaning on one side and the frequency of the genes technically labeled *ASPM* and *Microcephalin*, which affect the development of voice pitch.

The key properties of the mind guiding language evolution almost certainly appeared before the origin of language itself. Their wellsprings are thought to be in the earlier, more fundamental architecture of cognition. The flexibility in development of syntax has been documented in the variability of word orders in recently evolved creoles, pidgin, and sign languages, which are abundantly used on every continent. Granted that syntax may be skewed by early contact with conventional languages, such biasing influences can be discounted in at least one case, the sign language of the Al-Sayyid Bedouin. All of the members of this group live in the Negev region of Israel, and all are congenitally deaf. The group was founded two centuries ago by 150 individuals, and its members are descendants of two of the founder's five sons. All suffered profound prelingual hearing loss at all frequencies caused by a recessive gene on chromosome 13q12. As a result of inbreeding from that time forward, all of the 3,500 contemporary Al-Sayyid now share the condition. The community uses a sign language developed early in its history, employing independently derived word orders. These structures differ from those found both in the spoken languages in and around them and in other sign languages used in nearby communities.

The natural variability of grammar has been further illustrated by research in which the sequence of activities by people engaged in tasks were compared with the word order they used to describe the sequence. In one study, speakers of four languages (English, Turkish, Spanish, and Chinese) were asked to speak and also, separately, to reconstruct the event with the use of pictures. The same order of nonverbal communication (actor-patient-act, which is analogous to subject-object-verb of speech) turned out to be used by all the subjects. That, more or less, is the way people actually think through an action scenario. But it was less than fully consistent across the languages they used in speech. Actor-patient-act was the same as found in many languages of the world—and, most significantly, the newly developing gestural languages. So there does appear to be a biasing epigenetic rule for word order embedded in our deeper cognitive structure, but its final products in grammar are highly flexible and learned. So both Skinner and Chomsky appear to have been partly right, but Skinner more so.

The multiplicity of pathways in the evolution of elementary syntax suggests that few if any genetic rules guide the learning of language by individual human beings. The probable reason has been revealed in recent mathematical models of gene-culture evolution constructed by Nick Chater and his fellow cognitive scientists. It is simply that the rapidly changing environment of speech does not provide a stable environment for natural selection. Language varies too swiftly across generations and from one culture to the next for such evolution to occur. As a consequence, there is little reason to expect that the arbitrary properties of language, including the abstract syntactic principles of phrase structures and gene marking, have been built into a special "language module" of the brain by evolution. "The genetic basis of human language acquisition," the researchers conclude, "did not coevolve with language, but primarily predates the emergence of language. As suggested by Darwin, the fit between language and its underlying mechanisms arose because language has evolved to fit the human brain, rather than the reverse."

It is not going too far, I believe, to add that the failure of natural selection to create an independent universal grammar has played a major role in the diversification of culture and, from that flexibility and potential inventiveness, the flowering of human genius.

· 23 ·

The Evolution of Cultural Variation

GENE-CULTURE COEVOLUTION, THE impact of genes on culture and, reciprocally, culture on genes, is a process of equal importance to the natural sciences, the social sciences, and the humanities. Its study provides a way to connect these three great branches with a network of causal explanations.

If this claim seems overly bold, consider cultural variation among societies. It is commonly believed that if two societies have different culture traits in the same category—say, monogamy as opposed to polygamy, or warlike policies versus peaceable policies—then the evolutionary genesis of the patterns of variation and even the category itself must have been entirely cultural in nature, and genes had nothing to do with it.

This rush to judgment is due to an incomplete understanding of the relation between genes and culture. What genes prescribe or assist in prescribing is not one trait as opposed to another but the frequency of traits and the pattern they form as cultural innovation made them available. The expression of the genes may be plastic, allowing a society to choose one or more traits from among a multiplicity of choices. Or else it may *not* be plastic, allowing only one trait to be chosen by all societies.

Consider this familiar example of varying plasticity in anatomical traits. The genes prescribing the general development of fingerprints are very plastic in expression, allowing a vast number of variants among people. No two people in the world have completely identical fingerprints. In contrast, the genes prescribing the number of fingers on each hand are quite rigid. The number is five, always five. Only an extreme developmental accident or a mutation in the genes can yield another number.

The principle of varying plasticity is easily applied as well to cultural traits. The general practice of fashion in dress, ranging from loin cloth to white tie, has a genetic basis. However, because of the extreme (yet far from infinite) plasticity in the prescribing genes, and the multiple emotions they variously express, individuals select from several up to hundreds of options during their lifetimes. In another example, and at the opposite extreme, incest is instinctively avoided in all normal family settings owing to the Westermarck effect (very young children reared in close proximity are psychologically unable to bond sexually with each other at maturity).

Biologists who study development have discovered that the degree of plasticity in the expression of genes, like the presence or absence of the genes themselves, is subject to evolution by natural selection. It matters to the success of an individual whether he follows the dress fashion of his group and displays the correct insignia of his rank, occupation, and status. It mattered even more, to the point of life or death, in simpler societies of the kind formed during most of human evolution. In the case of the Westermarck effect, it has also mattered everywhere and under all circumstances, serving to provide all of humanity an automatic defense against the deadly effects of inbreeding.

All societies and each of the individuals in them play games of genetic fitness, the rules of which have been shaped across countless generations by gene-culture coevolution. When a rule is absolute, such as destruction by incest, there is only one hand to play; in this case, it is labeled "outbreed." When a part of the environment is unpredictable, on the other hand, the person is wise to use a mixed strategy achieved by plasticity. If one trait or response does not work, switch to another within the genetic repertory. The degree of plasticity existing within a category of culture depends not on any explicit judgment of what will occur in the future but on the degree of challenges to which the category of traits or behaviors had to respond in past generations when gene-culture coevolution was occurring.

Since the 1970s, biologists have been aware of the genetic processes by which the evolution of plasticity is most likely engineered. It is probably not by mutations of protein-coding genes, which prescribe a basic change in the amino acid composition of proteins. It is more likely by changes in the regulatory genes, which determine the rate and conditions under which the proteins are produced. Small changes in regulatory genes do not sound like much, but they can profoundly alter the proportions of anatomical structures and physiological activity. They can also target with greater precision certain parts of the body and particular physiological processes. Further, they can program sensitivity to selected stimuli impinging on the developing organism, with the result that different environments evoke the production of the particular variants best suited to live within them. Finally, mutations of regulatory genes, because they affect interactions in the developmental process, are less likely to be deleterious than mutations in protein-coding genes. They do not produce a new protein, and with that a structure or behavior built with the protein, a change that can easily perturb development in the remainder of the organism. Rather, they alter the amount of an existing protein, allowing finely tuned changes in a previous structure or behavior.

Ants and other social insects illustrate to an extreme degree the evolution of such adaptive plasticity. The workers of ant or termite colonies often differ so much from one another that they can easily be mistaken as belonging to separate species. Yet, in colonies with a single queen who mated with only one male, all the castes of a gender are close to being genetically identical. They are distinct in anatomy and behavior because as immature forms they were given either more food than others or less, leading to larger or smaller adults. While immature, their tissues also grew at different rates, so that larger and smaller individuals possessed different body proportions. The immatures were also sensitive to pheromones from adult colony numbers, altering the direction of development and how large they grew before reaching maturity. Researchers have documented still other factors that divide colony members into castes. Each caste specializes in its own labor role during its lifetime. One colony, with no significant genetic variation, can consist of virgin queens; small, timid minor workers; and giant soldiers with grotesquely enlarged heads and jaws.

In ants particularly, the elaboration of castes of ants out of plasticity is only part of a sophisticated process called "adaptive demography." Not only do the castes engage in specialized labor, but they are programmed to be created at a certain frequency in accordance with their natural death rate so as to produce ratios of castes optimal for the colony as a whole. For example, members of the large major caste of weaver ants, which conduct most of the work of the colony outside the nest, as well as defend the colony against enemies, have a higher death rate than minor workers, which serve as nurses inside the nest. As an evident consequence, the colony produces majors at a higher per capita rate than minors, maintaining what appears to be an optimum balance in numbers between the two castes.

Cultural variation in humans is determined mostly by two properties of social behavior, both of which are subject to evolution by natural selection. The first is the degree of bias in the epigenetic rule—very low in dress fashion, very high in incest avoidance. The second property of cultural variation is the likelihood that individual group members imitate others in the same society who have adapted the trait ("sensitivity to usage pattern").

To illustrate the solution of the gene-versus-culture conundrum, first note that the three rows of culture categories depicted in the accompanying figure differ from one another genetically. Choose one of the three, and take a point under each of the two nodes that have emerged (toward the bottom, owing to greater evolved tendency to imitate the actions of others). Let the points represent two societies. The two societies will likely have chosen different culture traits, even though they are genetically identical for the rules they follow in choosing. The properties are the epigenetic rules and the

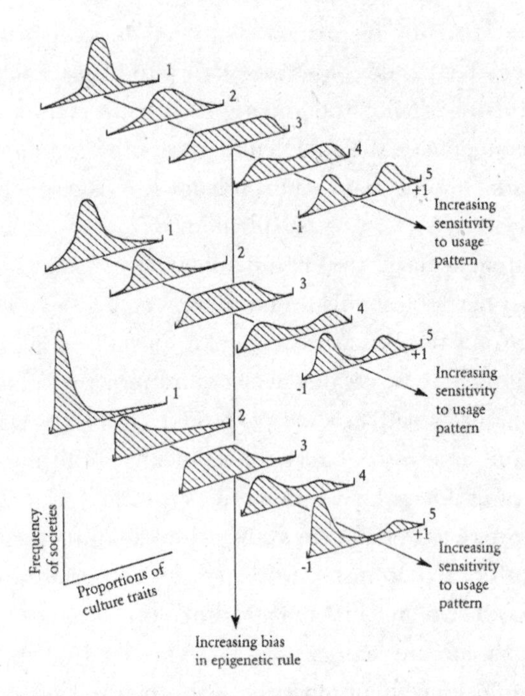

FIGURE 23-1. *The evolution of cultural variation, based on the simple case of two traits in the same category of culture (such as incest avoidance and dress fashion). The variation is measured as the number of societies choosing one of two traits in three categories of culture (top to bottom). The propensity to imitate others is interpreted to be sensitivity to usage by others. (Modified from a mathematical model by Charles J. Lumsden and Edward O. Wilson, "Translation of epigenetic rules of individual behavior into ethnographic patterns,"* Proceedings of the National Academy of Sciences U.S.A. *77[7]: 4382–4386 [1980]; also, Charles J. Lumsden and Edward O. Wilson,* Genes, Mind, and Culture: The Coevolutionary Process *[Cambridge, MA: Harvard University Press, 1981], p. 130.)*

propensity to imitate others, both of which have originated by gene-culture coevolution.

The intricacies of gene-culture coevolution are fundamental to understanding the human condition. They are complex and at first may seem strange, being unfamiliar. But with research employing the right measures and analysis, guided by evolutionary theory, they can be dissected into their essential elements.

· 24 ·

The Origins of Morality and Honor

ARE PEOPLE INNATELY GOOD, but corruptible by the forces of evil? Or, are they instead innately wicked, and redeemable only by the forces of good? People are both. And so it will forever be unless we change our genes, because the human dilemma was foreordained in the way our species evolved, and therefore an unchangeable part of human nature. Human beings and their social orders are intrinsically imperfectible and fortunately so. In a constantly changing world, we need the flexibility that only imperfection provides.

The dilemma of good and evil was created by multilevel selection, in which individual selection and group selection act together on the same individual but largely in opposition to each other. Individual selection is the result of competition for survival and reproduction among members of the same group. It shapes instincts in each member that are fundamentally selfish with reference to other members. In contrast, group selection consists of competition between societies, through both direct conflict and differential competence in exploiting the environment. Group selection shapes instincts that tend to make individuals altruistic toward one another (but not toward members of other groups). Individual selection is responsible for much of what we call sin, while group selection is responsible for the greater part of virtue. Together they have created the conflict between the poorer and the better angels of our nature.

Individual selection, defined precisely, is the differential longevity and fertility of individuals in competition with other members of the group. Group selection is differential longevity and lifetime fertility of those genes that prescribe traits of interaction among members of the group, having arisen during competition with other groups.

How to think out and deal with the eternal ferment generated by multilevel selection is the role of the social sciences and humanities. How to explain it is the role of the natural sciences, which, if successful, should make the pathways to harmony among the three great branches of learning easier to create. The social sciences and humanities are devoted to the proximate, outwardly expressed phenomena of human sensations and thought. In the same way that descriptive natural history is related to biology, the social sciences and humanities are related to human self-understanding. They describe how individuals feel and act, and with history and drama they tell a representative fraction of the infinite stories that human relationships can generate. All of this, however, exists within a box. It is confined there because sensations and thought are ruled by human nature, and human nature is also in a box. It is only one of a vast number of possible natures that could have evolved. The one we have is the result of the improbable pathway followed across millions of years by our genetic ancestors that finally produced us. To see human nature as the product of this evolutionary trajectory is to unlock the ultimate causes of our sensations and thought. To put together both proximate and ultimate causes is the key to self-understanding, the means to see ourselves as we truly are and then to explore outside the box.

In the search for ultimate causes of the human condition, the distinction between levels of natural selection applied to human behavior is not perfect. Selfish behavior, perhaps including nepotism-generating kin selection, can in some ways promote the interests of the group through invention and entrepreneurship. As the final touches of cognitive evolution were being added before and after the African breakout 60,000 years ago, there likely lived the equivalents of Medicis, Carnegies, and Rockefellers, who advanced themselves and their families in ways that also benefited their societies. Group selection in its turn promoted the genetic interests of individuals with privilege and status as rewards for outstanding performance on behalf of the tribe.

Nevertheless, an iron rule exists in genetic social evolution. It is that selfish individuals beat altruistic individuals, while groups of altruists beat groups of selfish individuals. The victory can never be complete; the balance of selection pressures cannot move to either extreme. If individual selection were to dominate, societies would dissolve. If group selection were to dominate, human groups would come to resemble ant colonies.

Each member of a society possesses genes whose products are targeted by individual selection and genes targeted by group selection. Each individual is linked to a network of other group members. Its own survival and reproductive capacity are dependent in part on its interaction with others in the network. Kinship influences the structure of the network, but it is not the key to its evolutionary dynamics, as is wrongly posited by inclusive-fitness theory. Instead, what counts is the hereditary propensity to form the myriad alliances, favors, exchanges of information, and betrayals that make up daily life in the network.

Throughout prehistory, as humanity evolved its cognitive prowess, the network of each individual was almost identical to that of the group to which he belonged. People lived in scattered bands of a hundred or fewer (thirty was probably a common number). They had knowledge of neighboring bands, and, judging from the lives of surviving hunter-gatherers, neighbors to some degree formed alliances. They participated in trade and exchanges of young women, but also in rivalries and vengeance raids. But the heart of each individual's social existence was the band, and the cohesion of the band was kept tight by the binding force of the network it composed.

With the emergence of villages and then chiefdoms in the Neolithic period around 10,000 years ago, the nature of the networks changed dramatically. They grew in size and broke into fragments. These subgroups became overlapping and at the same time hierar-

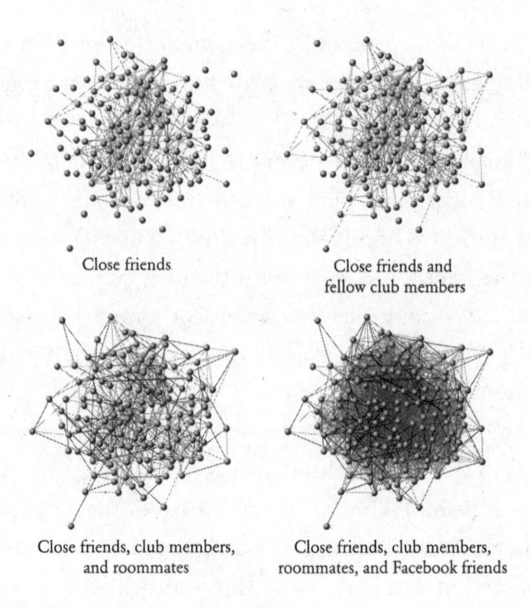

FIGURE 24-1. *In modern society, social networks such as those illustrated here in part for 140 university students, have grown much larger and more discordant than in prehistoric and earlier historic times. The internet revolution, producing arrangements such as Facebook, has recently catapulted the networks to a new level. (From Nicholas Christakis and James M. Fowler,* Connected: The Surprising Power of Our Social Networks *[New York: Little, Brown, 2009].)*

chical and porous. The individual lived in a kaleidoscope of family members, coreligionists, co-workers, friends, and strangers. His social existence became far less stable than the world of the hunter-gatherers. In modern industrialized countries, networks grew to a complexity that has proved bewildering to the Paleolithic mind we inherited. Our instincts still desire the tiny, united band-networks that prevailed during the hundreds of millennia preceding the dawn of history. Our instincts remain unprepared for civilization.

The trend has thrown confusion into the joining of groups, one of the most powerful human impulses. We are ruled by an urge—better, a compelling necessity—that began in our early primate ancestry. Every person is a compulsive group-seeker, hence an intensely tribal animal. He satisfies his need variously in an extended family, organized religion, ideology, ethnic group, or sports club, singly or in combination. The possibilities are vast. In each of our groups we find competition for status, but also trust and virtue, the signature products of group selection. We worry. We ask, to whom in this shifting global world of countless overlapping groups should we pledge our loyalty?

Through it all our instincts remain in command and confuse, but a few among them, if we obey them wisely, may save us. For example, we feel empathy. We stay our hand. A great deal of recent research has made it possible to see how the impulses of morality might work inside the brain. A promising start has been found in explaining the Golden Rule, which is perhaps the only precept found in all organized religions. The rule is fundamental to all moral reasoning. When the great theologian and philosopher Rabbi Hillel was challenged to explain the Torah in the time he could stand on one foot, he replied, "Do not do unto others that which is repugnant to you. All else is commentary."

The answer might equally well have been expressed as "coercive empathy," meaning that unless people are psychopaths, they automatically feel the pain of others. The brain, the neurobiologist Donald W. Pfaff argues in *The Neuroscience of Fair Play*, is an organ not merely divided into major parts but divided against itself. The primal fear triggered by stressful or anger-producing stimuli is a response becoming well understood at the molecular and cellular levels. It is counterbalanced by an automatic shutdown of fear-inducing thought when altruistic behavior is appropriate. Sliding toward hostile and potentially violent behavior, the individual "loses" himself psychologically. In the clash of emotions, he transfers his own identity a little bit to the other person.

The brain of our Janus-like species is a supremely complex system of intersecting nerve cells, hormones, and neurotransmitters. It creates processes that variously reinforce or cancel one another out, according to context.

Fear in part is a flow of impulses that pass through the amygdala, the almond-shaped structure in the brain containing connections to nerve-cell circuits that contribute, all at once, to fear, the memory of fear, and the suppression of fear. Signals traveling through these connections integrate and then travel to other parts of the forebrain and midbrain. It appears that while the emotions of fear come from the amygdala, more complex fearful thoughts about a particular person or object causing the emotion come from the information-processing centers of the cerebral cortex.

A second clue to the automatic nature of the suppression of fear and anger has been found in circuits of the anterior cingulate cortex and the insula, which help mediate the emotional response to the sensation of pain. The circuits affect not only the response to one's own pain but also the perception of another person's pain.

Pfaff is a distinguished scientist who is cautious about stringing together such fragments from recent brain research to create one big picture, but he has also seen the value of creating at least a plausible working theory about a phenomenon of such obvious importance to understanding human behavior. The blurring process built into the brain's circuitry, whether triggered by fear, mental stress, or other emotions, can account for a virtually endless repertoire of ethically acceptable behavioral choices. Pfaff provides an imaginary example to illustrate the process:

> The theory has four steps. In the *first* step, one person considers taking a certain action with regard to another; for example, Ms. Abbott considers knifing Mr. Besser in the stomach. Before the action takes place it is represented in the prospective actor's brain, as every act must be. This act will have consequences for the other individual that the would-be actor can understand, foresee, and remember. *Second*, Ms. Abbott envisions the target of this action, Mr. Besser. *Third* comes the crucial step: she *blurs* the difference between the other person and herself. Instead of seeing the consequences of her act for Mr. Besser, with gruesome effects to his guts and blood, she loses *the mental and emotional difference* between his blood and guts and her own. The *fourth*

step is the decision. Ms. Abbott is now less likely to attack Mr. Besser, because she shares his fear (or, more precisely, she shares in the fear he would experience if he knew what she was contemplating).

For the neuroscientist, this explanation of an ethical decision by the would-be knifer has one very attractive feature: it involves only the loss of information, not its effortful acquisition or storage. The learning of complex information and its storage in memory are deliberate, painstaking processes, but the *loss* of information seems to take place with no trouble at all. Damping any one of the many mechanisms involved in memory can explain the blurring of identity required by this theory. In the example of Ms. Abbott and Mr. Besser, as the result of a blurring of identity—a loss of individuality—the attacker temporarily puts herself in the other person's place. She avoids an unethical act because of shared fear.

Should this explanation of ethical decision-making hold up, it will find resonance in evolutionary biology's understanding of group selection. Human beings are prone to be moral—do the right thing, hold back, give aid to others, sometimes even at personal risk—because natural selection has favored those interactions of group members benefitting the group as a whole.

In addition to the origin of instinctive empathy, group selection can at least in part be invoked to explain cooperation, an even more important trait of human nature. In 2002 Ernst Fehr and Simon Gächter clearly framed the scientific problem as follows: "Human cooperation is an evolutionary puzzle. Unlike other creatures, people frequently cooperate with genetically unrelated strangers, often in large groups, with people they will never meet again, and when reproductive gains are small or absent. These patterns of cooperation cannot be explained with the evolutionary theory of kin selection and the selfish motives associated with signaling theory or the theory of reciprocal altruism."

Kin selection, as I have pointed out, cannot be the solution of the paradox. It might be thought to have worked in the bands of the early hunter-gatherers, where because of small numbers kinship of the members was close. But mathematical analyses have revealed that kin selection of itself is inoperable as an evolutionary dynamical force. When closely related individuals come together, such that cooperators are more likely to meet other genetic cooperators, the result will not by itself promote the origin of cooperation. Only group selection, with groups containing more cooperators pitted against groups with fewer cooperators, will result in a shift at the level of the species toward greater and wider instinctive cooperation.

During the first decade of this century, biologists and anthropologists focused intensely on the evolution of cooperation. What they concluded is that the phenomenon was achieved in human prehistory through a mix of innate responses. These responses include status seeking by individuals, the leveling of high status of individuals by the group, and the impulse to volunteer punishment and retribution for those who deviate too far from the norms of the group. Each of the behaviors contains elements of both selfishness and altruism. All are interlocked in cause and effect, and they originated by group selection.

The tangle of impulses created in the conscious brain have been finely cataloged by Steven Pinker in *The Blank Slate* (2002):

> The other-condemning emotions—contempt, anger, and disgust—prompt one to punish cheaters. The other-praising emotions—gratitude and an emotion that may be called elevation, moral awe, or being moved—prompt one to reward altruists. The other-suffering emotions—sympathy, compassion, and empathy—prompt one to help a needy beneficiary. And the self-conscious emotions—guilt, shame, and embarrassment—prompt one to avoid cheating or to repair its effects.

Relentless ambivalence and ambiguity are the fruits of the strange primate inheritance that rules the human mind. To be human is also to level others, especially those who appear to receive more than they have earned. Even within the ranks of the elite, delicate games are played to achieve ever higher status while steering through the successive ranks of jealous rivals. Be modest in demeanor, ever modest, is the necessary stratagem. This is a tricky business. As the seventeenth-century essayist François de La Rochefoucauld observed, "Modesty is due to a fear of incurring the well-merited envy and contempt which pursues those who are intoxicated by good fortune. It is a useless display of strength of mind; and the modesty of those who attain the highest eminence is due to a desire to appear even greater than their position."

It is also helpful to enhance reputation by what researchers have called indirect reciprocity, by which a reputation for altruism and cooperativeness accrues to an individual, even if the actions that build it are no more than ordinary. A saying in German exemplifies the tactic: *Tue Gutes und rede darüber.* Do good and talk about it. Doors are then opened, and opportunities for friendships and alliances increased.

Since everyone knows the game, people are always willing to counter it if they safely can. They are acutely sensitive to hypocrisy and ever ready to level those on the rise whose credentials are anything less than impeccable. All levelers, which means just about everybody, have a formidable armament at their disposal. Roasts, jokes, parodies, and mocking laughter are remedies to weaken the haughty and overly ambitious. The put-down is an art based on wit, the salt in the meal of conversation, as it has been called, in which excellence is to be treasured. One of the best known and arguably the most illustrious of all time is the response of Samuel Foote to John Montagu, fourth Earl of Sandwich, when warned that he would die either by venereal disease or by the hangman's noose. Foote responded, "My Lord, that will depend upon whether I embrace your lordship's mistress or your lordship's morals."

There is, of course, a great deal more to human cooperativeness than its efficiency and its protection by the dismantling of presumption. All normal people are capable of true altruism. We are unique among animals in the degree that we attend to the sick and injured, help the poor, comfort the bereaved, and even willingly risk our own lives to save strangers. Many, having helped others in an emergency, then leave without identifying themselves. Or if they stay, they devalue their heroism by an all but mandatory dismissal, "It was just my job" or "I only did what I would expect others to do for me."

Authentic altruism exists, as Samuel Bowles and other investigators have argued. It enhances the strength and competitiveness of groups, and it has been favored during human evolution by natural selection at the group level.

Additional studies suggest (but have not yet conclusively proved) that leveling is beneficial even for the most advanced modern societies. Those that do best for their citizens in quality of life, from education and medical care to crime control and collective self-esteem, also have the lowest income differential between the wealthiest and poorest citizens. Among twenty-three of the world's wealthiest countries and individual U.S. states, according to an analysis in 2009 by Richard Wilkinson and Kate Pickett, Japan, the Nordic countries, and the U.S. state of New Hampshire have both the narrowest wealth differential and the highest average quality of life. At the bottom are the United Kingdom, Portugal, and the remainder of the United States.

People gain visceral pleasure in more than just leveling and cooperating. They also enjoy seeing punishment meted out to those who do not cooperate (freeloaders, criminals) and even to those who do not contribute at levels commensurate with their status (the idle rich). The impulse to bring down the wicked is served in full measure by tabloid exposés and true-crime stories. It turns out that people not only passionately wish to see wrongdoers and layabouts punished; they are also willing to take part in administering justice—even at a cost to themselves. Scolding a fellow motorist who runs a red light, whistle-blowing on your employer, reporting an ongoing felony to police—many will perform such services even if they do not know the miscreants personally and risk paying a cost for their good citizenship, at the very least by loss of time.

In the brain, the administration of such "altruistic punishment" lights up the bilateral anterior insula, a center of the brain also activated by pain, anger, and disgust. Its payout is to society in greater order and less selfish draining of resources from the public commons. It does not come from a rational calculus on the part of the altruist. He may at first include in his ruminations the ultimate impact on himself and his kin. Authentic altruism is based on a biological instinct for the common good of the tribe, put in place by group selection, wherein groups of altruists in prehistoric times prevailed over groups of individuals in selfish disarray. Our species is not *Homo oeconomicus.* At the end of the day, it emerges as something more complicated and interesting. We are *Homo sapiens,* imperfect beings, soldiering on with conflicted impulses through an unpredictable, implacably threatening world, doing our best with what we have.

And beyond the ordinary instincts of altruism, there is something more, delicate and ephemeral in character but, when experienced, transformative. It is *honor,* a feeling born of innate empathy and cooperativeness. It is the final reserve of altruism that may yet save our race.

Honor is of course a two-edged sword. One side of the blade is devotion and sacrifice in war. These responses arise from the primal group instinct to confront and defend against an enemy seen as a threat to the group. The mood generated was captured perfectly by the young English poet Rupert Brooke in 1914, before the First World War fully unfolded in its unspeakable tragedy, and he was killed.

Blow, bugles, blow! They brought us, for our dearth,
Holiness, lacked so long, and Love, and Pain.
Honour has come back, as a king, to earth,
And paid his subjects with a royal wage;
And Nobleness walks in our ways again;
And we have come into our heritage.

The other edge of the sword is honor of the individual pitted against the crowd, and sometimes against a prevailing moral precept or even religion itself. It has been elegantly expressed by the philosopher Kwame Anthony Appiah in *The Honor Code: How Moral Revolutions Happen* (2010), in the following description of the resistance of individuals and minority resistance groups against organized injustice.

> You might ask what honor does in these stories that morality by itself does not. A grasp of morality will keep soldiers from abusing the human dignity of their prisoners. It will make them disapprove of the acts of those who don't. And it will allow women who have been vilely abused to know that their abusers deserve punishment. But it takes a sense of honor to drive a soldier beyond doing what is right and condemning what is wrong to insisting that something is done when others on his side do wicked things. It takes a sense of honor to feel implicated by the acts of others.
>
> And it takes a sense of your own dignity to insist, against the odds, on your right to justice in a society that rarely offers it to women like you; and a sense of the dignity of all women to respond to your own brutal rape not just with indignation and a desire for revenge but with a determination to remake your country, so that its women are treated with the respect you know they deserve. To make such choices is to live a life of difficulty; even, sometimes, of danger. It is also, and not incidentally, to live a life of honor.

The naturalistic understanding of morality does not lead to absolute precepts and sure judgments, but instead warns against basing them blindly on religious and ideological dogma. When such precepts are misguided, which is often, it is usually because they are based on ignorance. Some important factor or other was unintentionally omitted during the formulation. Consider, for example, the papal ban against artificial contraception. The decision was made—with good intentions—by one person, Paul VI in his 1968 encyclical *Humanae Vitae.* The reason he gave seems at first entirely reasonable. God, he posited, intends for sexual intercourse to be limited to the purpose of conceiving children. But the logic of *Humanae Vitae* is wrong. It leaves out a vital fact. An abundance of evidence from psychology and reproductive biology, much of it obtained since the 1960s, has revealed that there is another, additional purpose to sexual intercourse. Human females have hidden external genitalia and do not advertise estrus, thus differing from females of other primate species. Both men and women, when bonded, invite continuous and frequent intercourse. The practice is genetically adaptive: it ensures that the woman and her child have help from the father. For the woman, the commitment secured by pleasurable nonreproductive intercourse is important, even vital in many circumstances. Human infants, to acquire large organized brains and high intelligence, must go through an unusually long period of helplessness during their development. The mother cannot count on the same level of support from the community, even in tightly knit hunter-gatherer societies, that she obtains from a sexually and emotionally bonded mate.

A second example of dogmatic ethics gone wrong for lack of knowledge is homophobia. The basal reasoning is much the same as for opposition to artificial contraception: sex not intended for reproduction must be an aberration and a sin. But an abundance of evidence points to the opposite. Committed homosexuality, with the preference appearing in childhood, is heritable. This means the trait is not always fixed, but part of the greater likelihood of a person's developing into a homosexual is prescribed by genes that differ from those that lead to heterosexuality. It has further turned out that heredity-influenced homosexuality occurs in populations worldwide too frequently to be due to mutations alone. Population geneticists use a rule of thumb to account for abundance at this level: if a trait cannot be due solely to random mutations, and yet it lowers or eliminates reproduction in those who have it, then the trait must be favored by natural selection working on a target of some other kind. For example, a low dose of homosexual-tending genes may give competitive advantages to a practicing heterosexual.

Or, homosexuality may give advantages to the group by special talents, unusual qualities of personality, and the specialized roles and professions it generates. There is abundant evidence that such is the case in both preliterate and modern societies. Either way, societies are mistaken to disapprove of homosexuality because gays have different sexual preferences and reproduce less. Their presence should be valued instead for what they contribute constructively to human diversity. A society that condemns homosexuality harms itself.

There is a principle to be learned by studying the biological origins of moral reasoning. It is that outside the clearest ethical precepts, such as the condemnation of slavery, child abuse, and genocide, which all will agree should be opposed everywhere without exception, there is a larger gray domain inherently difficult to navigate. The declaration of ethical precepts and judgments made from them requires a full understanding of why we care about the matter one way or the other, and that includes the biological history of the emotions engaged. This inquiry has not been done. In fact, it is seldom even imagined.

With deepened self-understanding, how will we feel about morality and honor? I have no doubt that in many cases, perhaps the great majority, the precepts shared by most societies today will stand the test of biology-based realism. Others, such as the ban on artificial conception, condemnation of homosexual preference and forced marriages of adolescent girls, will not. Whatever the outcome, it seems clear that ethical philosophy will benefit from a reconstruction of its precepts based on both science and culture. If such greater understanding amounts to the "moral relativism" so fervently despised by the doctrinally righteous, so be it.

· 25 ·

The Origins of Religion

THE ARMAGEDDON IN THE CONFLICT between science and religion (if I may be allowed so strong a metaphor) began in earnest during the late twentieth century. It is the attempt by scientists to explain religion to its foundations—not as an independent reality within which humanity struggles to find its place, not as obeisance to a divine Presence, but as a product of evolution by natural selection. At its source, the struggle is not between people but between worldviews. People are not disposable, but worldviews are.

Was Man made in the image of God, or was God made in the image of Man? This is the heart of the difference between religion and science-based secularism. Which alternative is selected has profound importance for human self-understanding and the way people treat each other. If God made Man in His image, a belief suggested by the creation stories and iconographies of most religions, it is reasonable to suppose that He is personally in charge of humans. If, on the other hand, God did not create humanity in His image, then there is a good chance that the solar system is not special within the ten sextillion or so other star systems in the universe. If the latter alternative were widely suspected, devotion to organized religions would fall off significantly.

We then come to the ultimate question, which it seems to me theologians over the centuries have always complicated unnecessarily. Does God exist? If He does exist, is He a personal God, one to whom we may pray with the expectation of receiving an answer? And if that much is true, might we expect to be immortal, living, say, the next trillions of trillion years (just for a start) in peace and comfort?

On these basic questions a division widened during the twentieth century between religious believers and secular scientists. In 1910 a survey of "greater" (starred) scientists listed in *American Men of Science* revealed that a still sizable 32 percent believed in a personal God, and 37 percent believed in immortality. When the survey was repeated in 1933, believers in God had fallen to 13 percent and those in immortality to 15 percent. The trend continues. By 1998, members of the United States National Academy of Sciences, an elite elected group sponsored by the federal government, were approaching complete atheism. Only 10 percent testified to a belief in either God or immortality. Among them were a scant 2 percent of the biologists.

In modern civilizations, there is no overwhelming importance in the general populace to belong to an organized religion. Witness, for example, the strong differences in religiosity between people in the United States and those in western Europe. Polls published in the late 1990s found that more than 95 percent of Americans believed in God or some kind of universal life force, against 61 percent of the British. Eighty-four percent of Americans thought that Jesus is either God or the son of God, but only 46 percent of the British. In a poll taken in 1979, 70 percent of Americans believed in life after death, in contrast to 46 percent of the Italians, 43 percent of the French, and 35 percent of the Scandinavians. Nearly 45 percent of Americans today attend church more than once a week, compared with 13 percent of the British, 10 percent of the French, 3 percent of the Danish, and 2 percent of the Icelanders.

I am often asked the reason for these intercontinental disparities, given that most Americans are of western European descent. There is also considerable puzzlement over the widespread biblical literalism and denial, by half the U.S. population, of biological evolution. Having been raised as a Southern Baptist, an evangelical denomination that includes a large percentage of America's fundamentalist Christians, I know very well the power of the King James Bible, the warmth and generosity of those it unites, and the beleaguering they feel in a culture they view as turning increasingly godless. The incorruptible, unchallengeable Bible is the instrument of all spiritual needs. Its venerable passages are a bottomless well of meaning. In lonely moments believers find companionship, in grief they find comfort, and in moral errancy they expect redemption. "What a friend we have in Jesus," a favorite hymn intones. "All our sins and griefs to bear! What a privilege to carry everything to God in prayer!" There are historical reasons why fundamentalist Protestants make up such a large percentage of Americans, which I leave to historians to explain. But to those who believe that their culture might be broken by ridicule and reason, I say think again. There are circumstances under which intelligent, well-educated people equate their identity and the meaning of their lives with their religion, and this is one of them.

If a personal God, or gods, or nonmaterial spirits are not accepted at least to some degree, what of a divine force that created the universe? Might we all worship such a Creator—even if He has no special interest in us? This is the argument of deism, that material existence was begun with a purpose by something or someone. If so, the reason for the universe remains to this day undisclosed, 13.7 billion years after the big bang. A few serious scientists have argued that at the least there must be a creator God. The core of their reasoning is the anthropic principle, which holds that the laws of physics and their parameters had to be finely adjusted in order for star systems to evolve and for carbon-based life to evolve within them. Such is the ultimate Goldilocks universe that surrounds us in its physical entities and forces—not too little of this, not too much of that. For example, if the big bang had been a bit more powerful, matter would have been blown apart too fast for stars and planets to form. One has to admit that the anthropic principle is intriguing. However, as the historian Thomas Dixon expresses its difficulty,

> How do we know whether or not to be surprised by any given configuration of physical constants? Surely any combination is almost infinitely improbable? How, in any case, do we know that these constants are free to vary in the way these arguments assume they are, and not simply fixed by nature or linked to each others in a way we do not understand? And should the actual existence of trillions of other universes, as opposed to their merely possible existence, really make us any less surprised about the existence and physical make-up of our own (supposing we were surprised in the first place, which honestly I wasn't)?

This counterargument reflects the insight of Hume's Philo: "Having found in so many other subjects much more familiar, the imperfections and even contradictions of human reason, I never should expect any success from its feeble conjectures, in a subject so sublime, and so remote from the sphere of our observation."

Suppose, in contravention of this reasoning and by some means, we chose to interpret the physical laws of the universe as evidence of a supreme supernatural being. It would then be an enormous leap of faith to impute the biological history that unfolded on this planet to some divine intervention. If the evidence from biology and anthropology means anything, it would be another mistake of equal magnitude to envision, in the manner of Plato and Kant, universal ethical precepts that exist separate from the idiosyncrasies of human existence, hence the God-given moral law so eloquently posited by C. S. Lewis and other Christian apologists. There is every good reason instead to explain the origin of religion and morality as special events in the evolutionary history of humanity driven by natural selection.

The evidence that lies before us in great abundance points to organized religion as an expression of tribalism. Every religion teaches its adherents that they are a special fellowship and that their creation story, moral precepts, and privilege from divine power are superior to those claimed in other religions. Their charity and other acts of altruism are concentrated on their coreligionists;

when extended to outsiders, it is usually to proselytize and thereby strengthen the size of the tribe and its allies. No religious leader ever urges people to consider rival religions and choose the one they find best for their person and society. The conflict among religions is often instead an accelerant, if not a direct cause, of war. Devout believers value their faith above all else and are quick to anger if it is challenged. The power of organized religions is based upon their contribution to social order and personal security, not to the search for truth. The goal of religions is submission to the will and common good of the tribe.

The illogic of religions is not a weakness in them, but their essential strength. Acceptance of the bizarre creation myths binds the members together. Among the various prominent Christian denominations, we find the belief that those who have surrendered their will to Jesus will soon ascend bodily to heaven, and those left behind will suffer for a thousand years, after which the world will end. A rival faith disagrees, but recommends communion with Christ on Earth by eating his flesh and drinking his blood—both made literal by the act of transubstantiation. For outsiders openly to doubt such dogmas is regarded an invasion of privacy and a personal insult. For insiders to raise doubt is punishable heresy.

Such an intensely tribal instinct could, in the real world, arise in evolution only by group selection, tribe competing against tribe. The peculiar qualities of religious faith are the logical consequence of the dynamism at this higher level of biological organization.

The cores of traditional organized religions are their creation myths. How, in real-world history, did they originate? Some were drawn in part from folk memories of momentous events—of emigrations to new lands, of wars won or lost, of great floods and volcanic eruptions. Each was reworked and ritualized over generations. The perceived arrival of divine beings on the scene is made possible by the personal thought processes of the prophets and believers. They expect the gods to have the same emotions, reasoning, and motives as their own. In the Old Testament, for example, Yahweh was at different times loving, jealous, angry, and vengeful in the same manner as his mortal subjects.

People also project their humanness into animals, machines, places, and even fictional beings. It has been relatively easy in such transference to take the step from human rulers to invisible divine beings. For example, God in all three of the Abrahamic religions (Judaism, Christianity, and Islam) is a patriarch much like those in the desert kingdoms in which these religions arose.

Even the most phantasmagoric elements of creation myths—the appearance of demons and angels, voices of the unseen, the rise of the dead, and the halt of the sun in its orbit—are easy to understand not by physical laws but in the light of modern physiology and medicine. The clan leaders and shamans are always prone to talk with gods and spirits during dreams, drug-induced hallucinations, and bouts of mental illness. Especially vivid are episodes of night paralysis, during which otherwise healthy people step into an alternative world of threatening monsters and shattering fear. One subject studied by the psychologist J. Allan Cheyne describes "a shadow of a moving figure, arms outstretched, as [he] was absolutely sure it was supernatural and evil." Another was equally certain that he awoke to find the reality of "a half-snake/half-human thing shouting gibberish in [his] ear." The convincing imagery of sleep paralysis is closely similar to that of alien abductions, associated at least in some instances to hyperactivity in the parietal region of the brain. Other experiences reported during sleep paralysis include flying or falling, or leaving one's body. The primary emotion is fear, but that sometimes changes into excitement, exhilaration, or rapture.

Even more important in the creation of genesis myths are hallucinogenic drugs, which turn illusions into stories, longer in duration, full of symbols, and fraught with what the dreamer perceives as mystic significance. Shamans and their followers in primitive societies use them to connect with the spirit world. One such substance that has been especially well studied is ayahuasca, a hallucinogen taken widely among indigenous tribes of the Amazon Basin. To fall under the spell of ayahuasca is to experience vividly realistic visions, jumbled at first but then unfolding into some kind of a story. There variously appear odd geometric designs, jaguars, snakes, and other animals, and one's own death and journey to another world. One example is from a Siona Indian of Colombia who used *yagé*, the local name for ayahuasca:

> But then an aging woman came to wrap me in a great cloth, gave me to suckle at her breast, and then off I flew, very far, and suddenly I found myself in a completely illuminated place, very clear, where everything was placid and serene. There, where the yagé people live, like us, but better, is where one ends up.

Such might be interpreted as an entry to heaven. Next is a vision of hell, as experienced by a Chilean drug taker of European parentage. (Tigers refer to jaguars, the indigenous big cats of South America.)

> At first, many tiger faces. . . . Then *the* tiger. The largest and strongest of all. I know (for I read his thought) that I must follow him. I see the plateau. He walks with resolution in a straight line. I follow; but on reaching the edge and perceiving the brightness I cannot follow him.

She then looks into a circular pit of liquid fire, where people are swimming.

> The tiger wants me to go there. I don't know how to descend. I grasp the tiger's tail and he jumps. Because of his musculature the jump is graceful and slow. The tiger swims in the liquid fire as I sit on his back . . . I rise on the tiger on the shore . . . There is a crater. We wait for some time and there begins an enormous eruption. The tiger tells me I must throw myself into the crater . . .

These raw visions are no more bizarre than those posed as foundational truths by the world's major religions. We learn much of this in the testimony of Saint John the Divine in the New Testament's final chapter, the Book of Revelation. The year is the first century, probably AD 96, and the place the Greek island of Patmos. In Saint John's vision, Jesus returns to Earth from His throne in heaven at the right side of God and speaks through angels. John is startled by a strange voice.

> And I turned to see the voice that spake with me. And being turned, I saw seven golden candlesticks; and in the midst of the seven candlesticks one like unto the Son of Man, clothed with a garment down to the foot, and girt about the paps with a golden girdle. His head and his hairs *were* white like wool, as white as snow; and his eyes *were* as a flame of fire; and his feet like unto fine brass, as if they burned in a furnace; and his voice as the sound of many waters. And he had in his right hand seven stars: and out of his mouth went a sharp two-edged sword: and his countenance *was* as the sun shineth in his strength.

Jesus on this Second Coming (not the other catastrophic one He is about to promise John) is in an angry mood. He has mixed feelings about the seven cities represented by the candles, and He is disposed to strike down citizens in them who have wandered from their devotion to Him. He identifies himself as the Alpha and Omega, who holds the "keys of hell and death." Jesus especially hates the deeds of the Nicolaitans. And to the wayward church members in Patmos who also have gone over to the Nicolaitan doctrine, he issues a fierce warning, "Repent; or else I will come unto thee quickly, and will fight against thee with the sword of my mouth." Jesus, in Saint John's testimony, goes on through angels to foretell the Rapture, the Tribulation, and the war between the forces of God and Satan, ending in a final victory for God.

Saint John the Divine might have experienced a real divine visit just as he reported it. Far more likely, however, he had dreams from taking hallucinogenic drugs, still in his time a widespread practice in southeastern Europe and the Middle East. The most pow-

FIGURE 25-1. *Keeping the dead at home as well as in the spirit world. In a Kukukuku village of New Guinea, a dead elder mummified by smoke fire is surrounded by his family. (From Vernon Reynolds and Ralph Tanner,* The Biology of Religion *[New York: Longman, 1983].)*

erful used were made from deadly nightshade (*Atropa belladonna*), nightshade (species of *Datura*), ergot (*Claviceps purpurea*, a fungus that grows on grasses and sedges, and a source of LSD), and hemp (*Cannabis sativa*).

Just as likely, John could have been suffering from schizophrenia, which produces hallucinations similar to John's visions: voices, other sounds as conversations and commands—sometimes experienced as very forceful and important thoughts, often reassuring but at other times menacing. The delusions also expand into longer stories, and may coalesce into a fantasy-based worldview.

The case of Saint John the Divine is of more than ordinary importance, because the Book of Revelation, the climax and conclusion of the New Testament, serves as a guidebook for conservative evangelical Protestants. John's dreams have exercised a profound effect on the way millions of perfectly sane and responsible people view the world and to a varying extent order their lives. His declarations may be thought true, but, in my sober judgment the image of a baleful Jesus threatening to cleave dissidents with a first-century sword is so far out of line with the remainder of the New Testament as to make a simple biological explanation preferable.

In any case, historians and other scholars with an evolutionary perspective and undeterred by the supernatural assumptions of traditional theology, have begun to piece together the steps that led

FIGURE 25-2. *Seeking visions through self-torture. In the Mandan ritual Indian braves sought visions by having thongs inserted through their flesh and then being turned until they fainted. (From Vernon Reynolds and Ralph Tanner,* The Biology of Religion *[New York: Longman, 1983].)*

to the hierarchical and dogmatic structures of modern religions. At some point in Late Paleolithic times, people began to reflect on their own mortality. The earliest known burial sites with any sign of ritual are 95,000 years old. At that time, or before, the living must have asked, Where do all these dead people go? The answer would have been immediately obvious to them. The departed still lived, and regularly rejoined the living—in dreams. It was in the spirit world of dreams, and even more vividly in drug-induced hallucinations, that their deceased relatives dwelled, along with allies, enemies, gods, angels, demons, and monsters. Similar visions, as later societies found, could also be induced by fasting, exhaustion, and self-torture. Today, as then, the conscious mind of every living person leaves his body in sleep and enters the spirit world created by neuronal surges of his brain.

At some early time, shamans appeared and took charge of interpretation of the visions, particularly their own, which they deemed especially important. They asserted that the apparitions controlled the fate of the tribe. The supernatural beings were assumed to have

FIGURE 25-3. *Leader of the Mandan Buffalo Bull Society. (From Joseph Campbell, with Bill Moyers,* The Power of Myth *[New York: Doubleday, 1988]. Painting by Karl Bodmer, 1834.)*

the same emotions as living people, and for that reason they had to be honored and placated with ceremony. They had to be summoned to bless the little community during rites of passage—into adulthood, marriage, and death. With the Neolithic revolution, and especially during the emergence of states, when alliances were made for trade and war, and different tribes fought for religious supremacy, the gods were sometimes shared.

As social complexity grew, so did the responsibility of the gods for maintaining social stability, which their priestly human surrogates achieved by top-down political control. When political, military, and religious leaders collaborated to achieve these ends, dogma was both traditional and firm. When successful political revolutions occurred, religious leaders usually found a way to adjust to circumstance—typically by taking the side of the insurgents and softening the old establishment dogmas.

During the early Israelite formation of what was to become the powerful Abrahamic religions, there were still multiple gods presiding over the chosen people. In Psalms 86:8, the scribe intones, "Among the gods there is none like unto thee, O Lord; neither are there any works like unto thy works." In time, Yahweh gained absolute power over the Israelites. Thereafter, He tended to command tolerance toward the deities of neighboring kingdoms when times were good, and harsh oppression when times were hard.

Religious believers today, as in ancient times, are not as a rule

FIGURE 25-4. *Prehistoric and early historical dancers in mystical, animal-head disguise. (A) A Paleolithic cave painting from Trois Frères, France. (B) Prehistoric Bushman painting at Afvallingskop, South Africa. (C, D) Paintings by Sioux of the American Plains. (From R. Dale Guthrie,* The Nature of Paleolithic Art *[Chicago: University of Chicago Press, 2005].)*

much interested in theology, and not at all in the evolutionary steps that led to the present-day world religions. They are concerned instead with religious faith and the benefits it provides. The creation myths explain all they need to know of deep history in order to maintain tribal unity. In times of change and danger, their personal faith promises stability and peace. When faced by threat and competition from outside groups, the myths assure the believers that they are paramount in the sight of God. Religious faith offers the psychological security that uniquely comes from belonging to a group, and a divinely blessed one at that. At least within the immense throngs of Abrahamic faithful around the world, it promises eternal life after death, and in heaven, not hell—especially if we choose the right denomination within the many available, and pledge to faithfully practice its rituals.

All of the stimuli of awe and wonder, whose capacity is invested in the human mind, have been appropriated by religious faiths across centuries, in masterpieces of literature, the visual arts, music, and architecture. Three thousand years of Yahweh have wrought an aesthetic power in these creative arts second to none. There is nothing in my own experience more moving than the Roman Catholic

Lucernarium, when the *lumen Christi* (light of Christ) is spread by Paschal candlelight into a darkened cathedral; or the choral hymns to the standing faithful and approaching procession during an evangelical Protestant altar call.

These benefits require submission to God, or his Son the Redeemer, or both, or to His final chosen spokesman Muhammad. This is too easy. It is necessary only to submit, to bow down, to repeat the sacred oaths. Yet let us ask frankly, to whom is such obeisance really directed? Is it to an entity that may have no meaning within reach of the human mind—or may not even exist? Yes, perhaps it really is to God. But perhaps it is to no more than a tribe united by a creation myth. If the latter, religious faith is better interpreted as an unseen trap unavoidable during the biological history of our species. And if this is correct, surely there exist ways to find spiritual fulfillment without surrender and enslavement. Humankind deserves better.

· 26 ·

The Origins of the Creative Arts

RICH AND SEEMINGLY BOUNDLESS as the creative arts seem to be, each is filtered through the narrow biological channels of human cognition. Our sensory world, what we can learn unaided about reality external to our bodies, is pitifully small. Our vision is limited to a tiny segment of the electromagnetic spectrum, where wave frequencies in their fullness range from gamma radiation at the upper end, downward to the ultralow frequency used in some specialized forms of communication. We see only a tiny bit in the middle of the whole, which we refer to as the "visual spectrum." Our optical apparatus divides this accessible piece into the fuzzy divisions we call colors. Just beyond blue in frequency is ultraviolet, which insects can see but we cannot. Of the sound frequencies all around us we hear only a few. Bats orient with the echoes of ultrasound, at a frequency too high for our ears, and elephants communicate with grumbling at frequencies too low.

Tropical mormyrid fishes use electric pulses to orient and communicate in opaque murky water, having evolved to high efficiency a sensory modality entirely lacking in humans. Also, unfelt by us is Earth's magnetic field, which is used by some kinds of migratory birds for orientation. Nor can we see the polarization of sunlight from patches of the sky that honeybees employ on cloudy days to guide them from their hives to flower beds and back.

Our greatest weakness, however, is our pitifully small sense of taste and smell. Over 99 percent of all living species, from microorganisms to animals, rely on chemical senses to find their way through the environment. They have also perfected the capacity to communicate with one another with special chemicals called pheromones. In contrast, human beings, along with monkeys, apes, and birds, are among the rare life forms that are primarily audiovisual, and correspondingly weak in taste and smell. We are idiots compared with rattlesnakes and bloodhounds. Our poor ability to smell and taste is reflected in the small size of our chemosensory vocabularies, forcing us for the most part to fall back on similes and other forms of metaphor. A wine has a delicate bouquet, we say, its taste is full and somewhat fruity. A scent is like that of a rose, or pine, or rain newly fallen on the earth.

We are forced to stumble through our chemically challenged lives in a chemosensory biosphere, relying on sound and vision that evolved primarily for life in the trees. Only through science and technology has humanity penetrated the immense sensory worlds in the rest of the biosphere. With instrumentation, we are able to translate the sensory worlds of the rest of life into our own. And in the process, we have learned to see almost to the end of the universe, and estimated the time of its beginning. We will never orient by feeling Earth's magnetic field, or sing in pheromone, but we can bring all such information existing into our own little sensory realm.

By using this power in addition to examine human history, we can gain insights into the origin and nature of aesthetic judgment. For example, neurobiological monitoring, in particular measurements of the damping of alpha waves during perceptions of abstract designs, have shown that the brain is most aroused by patterns in which there is about a 20 percent redundancy of elements or, put roughly, the amount of complexity found in a simple maze, or two turns of a logarithmic spiral, or an asymmetric cross. It may be coincidence (although I think not) that about the same degree of complexity is shared by a great deal of the art in friezes, grillwork, colophons, logographs, and flag designs. It crops up again in the glyphs of the ancient Middle East and Mesoamerica, as well in the pictographs and letters of modern Asian languages. The same level

FIGURE 26-1. *Optical arousal in visual design. Of the three computer-generated figures, the one in the center, with an intermediate amount of complexity, is automatically the most stimulating. (Based on Gerda Smets,* Aesthetic Judgment and Arousal: An Experimental Contribution to Psycho-Aesthetics *[Leuven, Belgium: Leuven University Press, 1973].)*

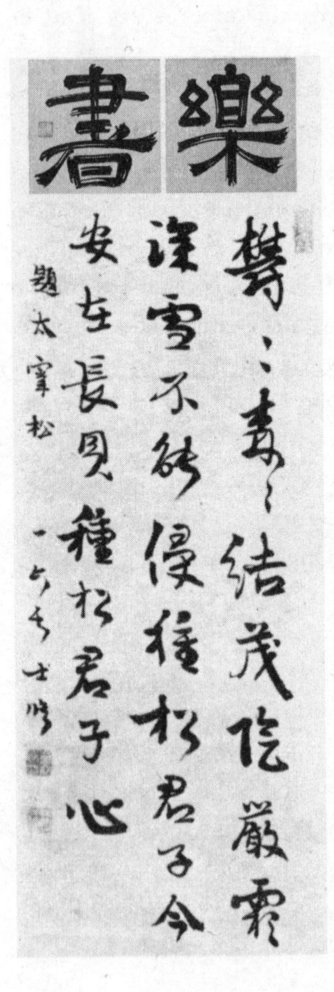

FIGURE 26-2. *The natural arousal by the complexity of Japanese pictographs is enhanced by the mood expressed through calligraphy. The two above are examples of* reisho *script, bold, linear, and simple, used in newspaper headings and on stone carvings. The one below is in* wayo *script, soft and elegant, used widely until the early twentieth century. (From Yūjirō Nakata,* The Art of Japanese Calligraphy *[New York: Weatherhill, 1973].)*

of complexity characterizes part of what is considered attractive in primitive art and modern abstract art and design. The source of the principle may be that this amount of complexity is the most that the brain can process in a single glance, in the same way that seven is the highest number of objects that can be counted at a single glance. When a picture is more complex, the eye grasps its content by the eye's sac-

ਜੈ ਘਰਿ ਕੀਰਤਿ ਆਖੀਐ ਕਰਤੇ ਕਾ ਹੋਇ ਬੀਚਾਰੋ ॥ ਤਿਤੁ ਘਰਿ ਗਾਵਹੁ ਸੋਹਿਲਾ ਸਿਵਰਿਹੁ ਸਿਰਜਨਹਾਰੋ॥੧॥ ਤੁਮ ਗਾਵਹੁ ਮੇਰੇ ਨਿਰਭਉ ਕਾ ਸੋਹਿਲਾ ॥ ਹਉਵਾਰੀ ਜਿਤੁ ਸੋਹਿਲੈ ਸਦਾ ਸੁਖੁ ਹੋਇ ॥੧॥ ਰਹਾਉ ॥ਨਿਤ ਨਿਤ ਜੀਅੜੇ ਸਮਾ-ਲੀਅਨਿ ਦੇਖੈਗਾ ਦੇਵਣਹਾਰੁ ॥ ਤੇਰੇ ਦਾਨੈ ਕੀਮਤਿ ਨਾ ਪਵੈ ਤਿਸੁ ਦਾਤੇਕਵਣੁਸੁਮਾਰੁ॥੨॥ਸੰਬਤਿਸਾਹਾ ਲਿਖਿਆ ਮਿਲਿ ਕਰਿ ਪਾਵਹੁ ਤੇਲ ॥ ਦੇਹੁ ਸਜਣ ਅਸੀਸੜੀਆ ਜਿਉ ਹੋਵੈ ਸਾਹਿਬ ਸਿਉ ਮੇਲੁ ॥੩॥ ਘਰਿ ਘਰਿ ਏਹੋ ਪਾਹੁਚਾ ਸਦੜੇ ਨਿਤ ਪਵੰਨਿ ॥ ਸਦਣਹਾਰਾ ਸਿਮਰੀਐ ਨਾਨਕ ਸੇ ਦਿਹ ਆਵੰਨਿ ॥੪॥੧॥

FIGURE 26-3. *The intrinsic beauty of Punjabi text, like that of many languages, is enhanced by the closeness of the symbols to the level of maximum automatic arousal. (From* Adi Granth, *the first computation of the Sikh scriptures, in Kenneth Katzner,* The Languages of the World, *new ed. [New York: Routledge, 1995].)*

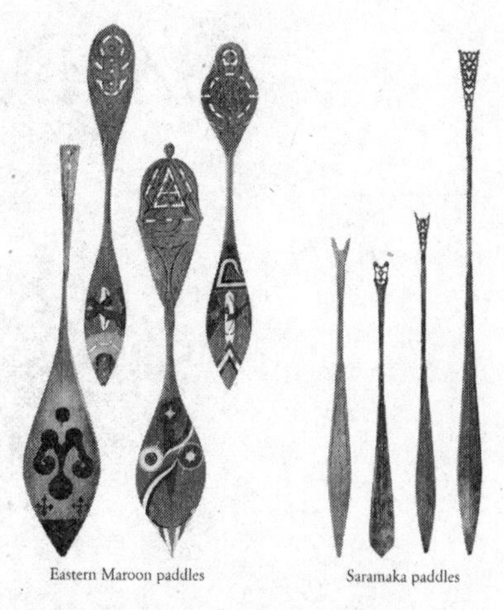

FIGURE 26-4. *The complexity of "primitive" art is typically close to that of maximum arousal. The paddles are the work of Surinamese villagers. (From Sally and Richard Price,* Afro-American Arts of the Suriname Rain Forest *[Berkeley: University of California Press, 1980].)*

cade or consciously reflective travel from one sector to the next. A quality of great art is its ability to guide attention from one of its parts to another in a manner that pleases, informs, and provokes.

In another sphere of the visual arts there is biophilia, the innate affiliation people seek with other organisms, and especially with the living natural world. Studies have shown that given freedom to choose the setting of their homes or offices, people across cultures gravitate toward an environment that combines three features, intuitively understood by landscape architects and real estate entrepreneurs. They want to be on a height looking down, they prefer open

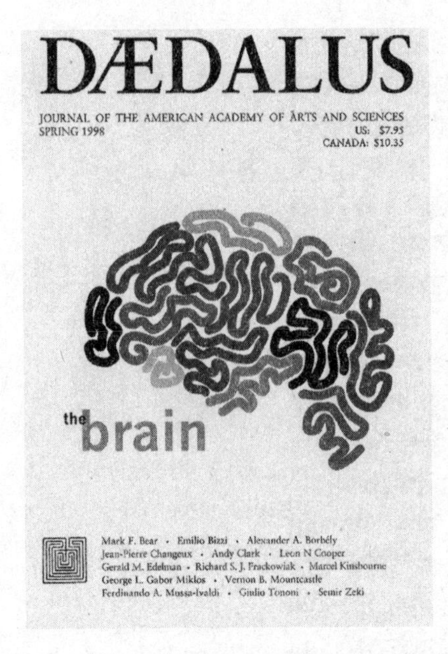

FIGURE 26-5. *Much of graphic art is composed of designs close to the level of automatic maximum arousal, as illustrated by the words, the central figure of the brain, and, in the lower left-hand corner, the symbol of the academic publisher. (Reproduced by permission of the American Academy of Arts and Sciences.)*

savanna-like terrain with scattered trees and copses, and they want to be close to a body of water, such as a river, lake, or ocean. Even if all these elements are purely aesthetic and not functional, home buyers will pay any affordable price to have such a view.

People, in other words, prefer to live in those environments in which our species evolved over millions of years in Africa. Instinctively, they gravitate toward savanna forest (parkland) and transitional forest, looking out safely over a distance toward reliable sources of food and water. This is by no means an odd connection, if considered as a biological phenomenon. All mobile animal species are guided by instincts that lead them to habitats in which they have a maximum chance for survival and reproduction. It should come as no surprise that during the relatively short span since the

FIGURE 26-6. *The habitation innately preferred by people has had a significant impact on landscape architecture. Believed by many researchers to have originated during prehuman evolution in the African savanna forest, the predilection includes dwelling on a height that is near a body of water and looks down on fruitful parkland (with large animals in sight, even if only represented by sculpture). This example is at the Deere Company headquarters at Moline, Illinois. (From* Modern Landscape Architecture: Redefining the Garden *[New York: Abbeville Press, 1991]. Photography by Felice Frankel, text by Jory Johnson.)*

beginning of the Neolithic, humanity still feels a residue of that ancient need.

If ever there was a reason for bringing the humanities and science closer together, it is the need to understand the true nature of the human sensory world, as contrasted with that seen by the rest of life. But there is another, even more important reason to move toward consilience among the great branches of learning. Substantial evidence now exists that human social behavior arose genetically by multilevel evolution. If this interpretation is correct, and a growing number of evolutionary biologists and anthropologists believe it is, we can expect a continuing conflict between components of behavior favored by individual selection and those favored by group selection. Selection at the individual level tends to create competitiveness and selfish behavior among group members—in status, mating, and the securing of resources. In opposition, selection between groups tends to create selfless behavior, expressed in greater generosity and altruism, which in turn promote stronger cohesion and strength of the group as a whole.

An inevitable result of the mutually offsetting forces of multilevel selection is permanent ambiguity in the individual human mind, leading to countless scenarios among people in the way they bond, love, affiliate, betray, share, sacrifice, steal, deceive, redeem, punish, appeal, and adjudicate. The struggle endemic to each person's brain, mirrored in the vast superstructure of cultural evolution, is the fountainhead of the humanities. A Shakespeare in the world of ants, untroubled by any such war between honor and treachery, and chained by the rigid commands of instinct to a tiny repertory of feeling, would be able to write only one drama of triumph and one of tragedy. Ordinary people, on the other hand, can invent an endless variety of such stories, and compose an infinite symphony of ambience and mood.

What exactly, then, are the humanities? An earnest effort to define them is to be found in the U.S. congressional statute of 1965, which established the National Endowment for the Humanities and the National Endowment for the Arts:

> The term "humanities" includes, but is not limited to, the study of the following: language, both modern and classical; linguistics; literature; history; jurisprudence; philosophy; archaeology; comparative religion; ethics; the history, criticism, and theory of the arts; those aspects of social sciences which have humanistic content and employ humanistic methods; and the study and application of the humanities to the human environment with particular attention to reflecting our diverse heritage, traditions, and history and to the relevance of the humanities to the current conditions of national life.

Such may be the scope of the humanities, but it makes no allusion to the understanding of the cognitive processes that bind them all together, nor their relation to hereditary human nature, nor their origin in prehistory. Surely we will never see a full maturing of the humanities until these dimensions are added.

Since the fading of the original Enlightenment during the late eighteenth and early nineteenth centuries, stubborn impasse has existed in the consilience of the humanities and natural sciences. One way to break it is to collate the creative process and writing styles of literature and scientific research. This might not prove so difficult as it first seems. Innovators in both of two domains are basically dreamers and storytellers. In the early stages of creation of both art and science, everything in the mind is a story. There is an imagined denouement, and perhaps a start, and a selection of bits and pieces that might fit in between. In works of literature and science alike, any part can be changed, causing a ripple among the other parts, some of which are discarded and new ones added. The surviving fragments are variously joined and separated, and moved about as the story forms. One scenario emerges, then another. The scenarios, whether literary or scientific in nature, compete. Words and sentences (or equations or experiments) are tried. Early on an end to all the imagining is conceived. It seems a wondrous denouement (or scientific breakthrough). But is it the best, is it true? To bring the end safely home is the goal of the creative mind. Whatever that might be, wherever located, however expressed, it begins as a phantom that might up until the last moment fade and be replaced. Inexpressible thoughts flit along the edges. As the best fragments solidify, they are put in place and moved about, and the story grows and reaches its inspired end. Flannery O'Connor asked, correctly, for all of us, literary authors and scientists, "How can I know what I mean until I see what I say?" The novelist says, "Does that work?," and the scientist says, "Could that possibly be true?"

The successful scientist thinks like a poet but works like a bookkeeper. He writes for peer review in hopes that "statured" scientists, those with achievements and reputations of their own, will accept his discoveries. Science grows in a manner not well appreciated by nonscientists: it is guided as much by peer approval as by the truth of its technical claims. Reputation is the silver and gold of scientific careers. Scientists could say, as did James Cagney upon receiving an Academy Award for lifetime achievement, "In this business you're only as good as the other fellow thinks you are."

But in the long term, a scientific reputation will endure or fall upon credit for authentic discoveries. The conclusions will be tested repeatedly, and they must hold true. Data must not be questionable, or theories crumble. Mistakes uncovered by others can cause a reputation to wither. The punishment for fraud is nothing less than death—to the reputation, and to the possibility of further career advancement. The equivalent capital crime in literature is plagiarism. But not fraud! In fiction, as in the other creative arts, a free play of imagination is expected. And to the extent it proves aesthetically pleasing, or otherwise evocative, it is celebrated.

The essential difference between literary and scientific style is the use of metaphor. In scientific reports, metaphor is permissible—provided it is chaste, perhaps with just a touch of irony and self-deprecation. For example, the following would be permitted in the introduction or discussion of a technical report: "This result if confirmed will, we believe, open the door to a range of further fruitful investigations." Not permitted is: "We envision this result, which we found extraordinarily hard to obtain, to be a potential watershed from which many streams of new research will surely flow."

What counts in science is the importance of the discovery. What matters in literature is the originality and power of the metaphor. Scientific reports add a tested fragment to our knowledge of the material world. Lyrical expression in literature, on the other hand, is a device to communicate emotional feeling directly from the mind of the writer to the mind of the reader. There is no such goal in scientific reporting, where the purpose of the author is to persuade the reader by evidence and reasoning of the validity and importance of the discovery. In fiction the stronger the desire to share emotion, the more lyrical the language must be. At the extreme, the statement may be obviously false, because author and reader want it that way. To the poet the sun rises in the east and sets in the west, tracking our diel cycles of activity, symbolizing birth, the high noon of life, death, and rebirth—even though the sun makes no such movement. It is just the way our distant ancestors visualized the celestial sphere and the starry sky. They linked its mysteries, which were many, to those in their own lives, and wrote them down in sacred script and poetry across the ages. It will be a long time before a similar venerability in literature is acquired by the real solar system, in which Earth is a spinning planet encircling a minor star.

On behalf of this other truth, that special truth sought in literature, E. L. Doctorow asks,

> Who would give up the *Iliad* for the "real" historical record? Of course the writer has a responsibility, whether as solemn interpreter or satirist, to make a composition that serves a revealed truth. But we demand that of all creative artists, of whatever medium. Besides which a reader of fiction who finds, in a novel, a familiar public figure saying and doing things not reported elsewhere knows he is reading fiction. He knows the novelist hopes to lie his way to a greater truth than is possible with factual reportage. The novel is an aesthetic rendering that would portray a public figure interpretively no less than the portrait on an easel. The novel is not read as a newspaper is read; it is read as it is written, in the spirit of freedom.

Picasso expressed the same idea summarily: "Art is the lie that helps us to see the truth."

The creative arts became possible as an evolutionary advance when humans developed the capacity for abstract thought. The human mind could then form a template of a shape, or a kind of object, or an action, and pass a concrete representation of the conception to another mind. Thus was first born true, productive language, constructed from arbitrary words and symbols. Language was followed by visual art, music, dance, and the ceremonies and rituals of religion.

The exact date at which the process leading to authentic creative arts is unknown. As early as 1.7 million years ago, ancestors of modern humans, most likely *Homo erectus*, were shaping crude teardrop-shaped stone tools. Held in the hand, they were probably used to chop up vegetables and meat. Whether they were also held in the mind as a mental abstraction, rather than merely created by imitation among group members, is unknown.

By 500,000 years ago, in the time of the much brainier *Homo heidelbergensis*, a species intermediate in age and anatomy between *Homo erectus* and *Homo sapiens*, the hand axes had become more sophisticated, and they were joined by carefully crafted stone blades and projectile points. Within another 100,000 years, people were using wooden spears, which must have taken several days and multiple steps to construct. In this period, the Middle Stone Age, the human ancestors began to evolve a technology based on a true, abstraction-based culture.

Next came pierced snail shells thought to be used as necklaces, along with still more sophisticated tools, including well-designed bone points. Most intriguing are engraved pieces of ocher. One design, 77,000 years old, consists of three scratched lines that connect a row of nine X-shaped marks. The meaning, if any, is unknown, but the abstract nature of the pattern seems clear.

Burials began at least 95,000 years ago, as evidenced by thirty individuals excavated at Qafzeh Cave in Israel. One of the dead, a nine-year-old child, was positioned with its legs bent and a deer antler in its arms. That arrangement alone suggests not just an abstract awareness of death but also some form of existential anxiety. Among today's hunter-gatherers, death is an event managed by ceremony and art.

The beginnings of the creative arts as they are practiced today may stay forever hidden. Yet they were sufficiently established by genetic and cultural evolution for the "creative explosion" that began approximately 35,000 years ago in Europe. From this time on until the Late Paleolithic period over 20,000 years later, cave art flourished. Thousands of figures, mostly of large game animals, have been found in more than two hundred caves distributed through southwestern France and northeastern Spain, on both sides of the Pyrenees. Along with cliffside drawings in other parts of the world, they present a stunning snapshot of life just before the dawn of civilization.

The Louvre of the Paleolithic galleries is at the Grotte Chauvet in the Ardèche region of southern France. The masterpiece among its productions, created by a single artist with red ocher, charcoal, and engraving, is a herd of four horses (a native wild species in Europe at that time) running together. Each of the animals is represented by only its head, but each is individual in character. The herd is tight and oriented obliquely, as though seen from slightly above and to the left. The edges of the muzzles were chiseled into bas relief to bring them into greater prominence. Exact analyses of the figures have found that multiple artists first painted a pair of rhinoceros males in head-to-head combat, then two aurochs (wild cattle) facing away. The two groups were placed to leave a space in the middle. Into the space the single artist stepped to create his little herd of horses.

The rhinos and cattle have been dated to 32,000–30,000 years before the present, and the assumption has been that the horses are that old as well. But the elegance and technology evident in the horses have led some experts to reckon their provenance as dating to the Magdalenian period, which extended from 17,000 to 12,000 years ago. That would align the origin with the great works on the cave walls of Lascaux in France and Altamira in Spain.

Apart from the exact date of the Chauvet herd's antiquity, the important function of the cave art remains uncertain. There is no reason to suppose the caves served as proto-churches, in which bands gathered to pray to the gods. The floors are covered with the remains of hearths, bones of animals, and other evidences of long-term domestic occupation. The first *Homo sapiens* entered central and eastern Europe around 45,000 years ago. Caves in that period obviously served as shelters that allowed people to endure harsh winters on the Mammoth Steppe, the great expanse of grassland that extended below the continental ice sheet across the whole of Eurasia and into the New World.

Perhaps, some writers have argued, the cave paintings were made to conjure sympathetic magic and increase the success of hunters in the field. This supposition is supported by the fact that a great majority of the subjects are large animals. Furthermore, 15 percent of these animal paintings depict animals that have been wounded by spears or arrows.

Additional evidence of a ritualistic content in the European cave art has been provided by the discovery of a painting of what is most likely a shaman with a deer headdress, or possibly a real deer's head. Also preserved are sculptures of three "lion-men," with human bodies and the heads of lions—precursors of the chimeric half-animal-half-gods later to show up in the early history of the Middle East. Admittedly, we have no testable idea of what the shaman did or the lion-men represented.

A contrary view of the role of cave art has been advanced by the wildlife biologist R. Dale Guthrie, whose masterwork *The Nature of Paleolithic Art* is the most thorough on the subject ever published. Almost all of the art, Guthrie argues, can be explained as the representations of everyday Aurignacian and Magdalenian life. The animals depicted belong to the species the cave dwellers regularly hunted (with a few, like lions, that may have hunted people), so naturally that would be a regular subject for talk and visual communication. There were also more figures of humans or at least parts of the human anatomy that are usually not mentioned in accounts of cave art. These tend to be pedestrian. The inhabitants often made prints by holding their hands on the wall and spewing ocher powder from their mouths, leaving an outline of spread thumb and fingers behind. The size of the hands indicates that it was mostly children who engaged in this activity. A good many graffiti are present as well, with meaningless squiggles and crude representations of male and female genitalia common among them. Sculptures of grotesque obese women are also present and may have been offerings to the spirits or gods to increase fertility—the little bands needed all the members they could generate. On the other hand, the sculptures might as easily have been an exaggerated representation of the plumpness in women desired during the frequent hard times of winter on the Mammoth Steppe.

The utilitarian theory of cave art, that the paintings and scratchings depict ordinary life, is almost certainly partly correct, but not entirely so. Few experts have taken into account that there also occurred, in another wholly different domain, the origin and use of music. This event provides independent evidence that at least some of the paintings and sculptures did have a magical content in the lives of the cave dwellers. A few writers have argued that music had no Darwinian significance, that it sprang from language as a pleasant "auditory cheesecake," as one author once put it. It is true that scant evidence exists of the content of the music itself—just as, remarkably, we have no score and therefore no record of Greek and Roman music, only the instruments. But musical instruments also existed from an early period of the creative explosion. "Flutes," technically better classified as pipes, fashioned from bird bones, have been found that date to 30,000 years or more before the present. At Isturitz in France and other localities some 225 reputed pipes have been so classified, some of which are of certain authenticity. The best among them have finger holes set in an oblique alignment and rotated clockwise to a degree seemingly meant to line up with the fingers of a human hand. The holes are also beveled in a way that allows the tips of the fingers to be sealed against them. A modern flutist, Graeme Lawson, has played a replica made from one of them, albeit of course without a Paleolithic score in hand.

Other artifacts have been found that can plausibly be interpreted as musical instruments. They include thin flint blades that, when hung together and struck, produce pleasant sounds like those from wind chimes. Further, although perhaps just a coincidence, the sections of walls on which cave paintings were made tend to emit arresting echoes of sound in their vicinity.

Was music Darwinian? Did it have survival value for the Paleolithic tribes that practiced it? Examining the customs of contemporary hunter-gatherer cultures from around the world, one can hardly come to any other conclusion. Songs, usually accompanied by dances, are all but universal. And because Australian aboriginals have been isolated since the arrival of their forebears about 45,000 years ago, and their songs and dances are similar in genre to those of other hunter-gatherer cultures, it is reasonable to suppose that they resemble the ones practiced by their Paleolithic ancestors.

Anthropologists have paid relatively little attention to contemporary hunter-gatherer music, relegating its study to specialists on music, as they are also prone to do for linguistics and ethnobotany (the study of plants used by the tribes). Nonetheless, songs and dances are major elements of all hunter-gatherer societies. Furthermore, they are typically communal, and they address an impressive array of life issues. The songs of the well-studied Inuit, Gabon pygmies, and Arnhem Land aboriginals approach a level of detail and sophistication comparable to those of advanced modern civilizations. The musical compositions of modern hunter-gatherers generally serve basically as tools that invigorate their lives. The subjects within the repertoires include histories and mythologies of the tribe as well as practical knowledge about land, plants, and animals.

Of special importance to the meaning of game animals in the Paleolithic cave art of Europe, the songs and dances of the modern tribes are mostly about hunting. They speak of the various prey; they empower the hunting weapons, including the dogs; they appease the animals they have killed or are about to kill; and they offer homage to the land on which they hunt. They recall and celebrate successful hunts of the past. They honor the dead and ask the favor of the spirits who rule their fates.

It is self-evident that the songs and dances of contemporary hunter-gatherer peoples serve them at both the individual and the group levels. They draw the tribal members together, creating a common knowledge and purpose. They excite passion for action. They are mnemonic, stirring and adding to the memory of information that serves the tribal purpose. Not least, knowledge of the songs and dances gives power to those within the tribe who know them best.

To create and perform music is a human instinct. It is one of the true universals of our species. To take an extreme example, the neuroscientist Aniruddh D. Patel points to the Pirahã, a small tribe in the Brazilian Amazon: "Members of this culture speak a language without numbers or a concept of counting. Their language has no fixed terms for colors. They have no creation myths, and they do not draw, aside from simple stick figures. Yet they have music in abundance, in the form of songs."

Patel has referred to music as a "transformative technology." To the same degree as literacy and language itself, it has changed the way people see the world. Learning to play a musical instrument even alters the structure of the brain, from subcortical circuits that encode sound patterns to neural fibers that connect the two cerebral hemispheres and patterns of gray matter density in certain regions of the cerebral cortex. Music is powerful in its impact on human feeling and on the interpretation of events. It is extraordinarily complex in the neural circuits it employs, appearing to elicit emotion in at least six different brain mechanisms.

Music is closely linked to language in mental development and in some ways appears to be derived from language. The discrimination patterns of melodic ups and downs are similar. But whereas language acquisition in children is fast and largely autonomous, music is acquired more slowly and depends on substantial teaching and practice. There is, moreover, a distinct critical period for learning language during which skills are picked up swiftly and with ease, whereas no such sensitive period is yet known for music. Still, both language and music are syntactical, being arranged as discrete elements—words, notes, and chords. Among persons with congenital defects in perception of music (composing 2 to 4 percent of the population), some 30 percent also suffer disability in pitch contour, a property shared in parallel manner with speech.

Altogether, there is reason to believe that music is a newcomer in human evolution. It might well have arisen as a spin-off of speech. Yet, to assume that much is not also to conclude that music is merely a cultural elaboration of speech. It has at least one feature not shared with speech—beat, which in addition can be synchronized from song to dance.

It is tempting to think that the neural processing of language served a preadaptation to music, and that once music originated it proved sufficiently advantageous to acquire its own genetic predisposition. This is a subject that will greatly reward deeper additional research, including the synthesis of elements from anthropology, psychology, neuroscience, and evolutionary biology.

VI

Where Are We Going?

· 27 ·

A New Enlightenment

SCIENTIFIC KNOWLEDGE and technology double every one to two decades, depending on the discipline in which information is measured. This exponential growth makes the future impossible to predict beyond a decade, let alone centuries or millennia. Futurists are therefore prone to dwell upon those directions which, in their opinion, humanity should go. But given our miserable lack of self-understanding as a species, the better goal at this time may be to choose where *not* to go. What, then, should we be careful to avoid? In thinking about the subject, we are destined always to come back full circle to the existential questions—Where do we come from? What are we? Where are we going?

Human beings are actors in a story. We are the growing point of an unfinished epic. The answer to the existential questions must lie in history, and that, of course, is the approach taken by the humanities. But conventional history by itself is truncated, in both its timeline and its perception of the human organism. History makes no sense without prehistory, and prehistory makes no sense without biology.

Humanity is a biological species in a biological world. In every function of our bodies and mind and at every level, we are exquisitely well adapted to live on this particular planet. We belong in the biosphere of our birth. Although exalted in many ways, we remain an animal species of the global fauna. Our lives are restrained by the two laws of biology: all of life's entities and processes are obedient to the laws of physics and chemistry; and all of life's entities and processes have arisen through evolution by natural selection.

The more we learn about our physical existence, the more apparent it becomes that even the most complex forms of human behavior are ultimately biological. They display the specializations evolved across millions of years by our primate ancestors. The indelible stamp of evolution is clear in the idiosyncratic manner in which humanity's sensory channels narrow our unaided perception of reality. It is confirmed in the way hereditarily prepared and counter-prepared programs guide the development of mind.

Still, we cannot escape the question of free will, which some philosophers still argue sets us apart. It is a product of the subconscious decision-making center of the brain that gives the cerebral cortex the illusion of independent action. The more the physical processes of consciousness have been defined by scientific research, the less has been left to any phenomenon that can be intuitively labeled as free will. We are free as independent beings, but our decisions are not free of all the organic processes that created our personal brains and minds. Free will therefore appears to be ultimately biological.

Yet, by any conceivable standard, humanity is far and away life's greatest achievement. We are the mind of the biosphere, the solar system, and—who can say?—perhaps the galaxy. Looking about us, we have learned to translate into our narrow audiovisual systems the sensory modalities of other organisms. We know much of the physicochemical basis of our own biology. We will soon create simple organisms in the laboratory. We have learned the history of the universe and look out almost to its edge.

Our ancestors were one of only two dozen or so animal lines ever to evolve eusociality, the next major level of biological organization above the organismic. There, group members across two or more generations stay together, cooperate, care for the young, and divide labor in a way favoring reproduction of some individuals over that in others. The prehumans were far greater in physical size than any of the eusocial insects and other invertebrates. They were endowed with much larger brains from the start. In time they hit upon the symbol-based language, and literacy, and science-based technology that give us the edge over the rest of life. Now, except for behaving like apes much of the time and suffering genetically limited life spans, we are godlike.

What dynamical force lifted us to this high estate? That is a question of enormous importance for self-understanding. The apparent answer is multilevel natural selection. At the higher level of the two relevant levels of biological organization, groups compete with groups, favoring cooperative social traits among members of the same group. At the lower level, members of the same group compete with one another in a manner that leads to self-serving behavior. The opposition between the two levels of natural selection has resulted in a chimeric genotype in each person. It renders each of us part saint and part sinner.

The interpretation of human selection forces I have presented in *The Social Conquest of Earth*, on the basis of recent research, opposes the theory of inclusive fitness and replaces it with standard models of population genetics applied to multiple levels of natural selection. Inclusive fitness is based on kin selection, in which individuals tend to cooperate with one another, or not, according to how close they are genealogically. This mode of selection, if defined broadly enough, was thought to explain all forms of social behavior, including advanced social organization. The opposing explanation, including a mathematical critique of inclusive-fitness theory, was fully developed from 2004 to 2010.

Given the technical complexity and importance of the subject, the controversy engendered by the new approach can be expected to continue for years, perhaps long after my own ability to grasp new data comes to an end. In the event, however, that the theory of inclusive fitness continues to be widely used, that should have little effect on the perception of group selection as the driving force of where we have been and where we are going. Theorists of inclusive fitness themselves have argued that kin selection can be translated into group selection, even though that belief now has been disproved mathematically. More importantly, group selection is clearly the process responsible for advanced social behavior. It also possesses the two elements necessary for evolution. First, group-level traits, including cooperativeness, empathy, and patterns of networking, have been found to be heritable in humans—that is, they vary genetically in some degree from one person to the next. And second, cooperation and unity manifestly affect the survival of groups that are competing.

It is further the case that the perception of group selection as the main driving force of evolution fits well with a great deal of what is most typical—and perplexing—about human nature. It also finds resonance in the evidence from the otherwise disparate fields of social psychology, archaeology, and evolutionary biology that human beings are intensely tribalist by nature. A basic element of human nature is that people feel compelled to belong to groups and, having joined, consider them superior to competing groups.

Multilevel selection (group and individual selection combined) also explains the conflicted nature of motivations. Every normal person feels the pull of conscience, of heroism against cowardice, of truth against deception, of commitment against withdrawal. It is our fate to be tormented with large and small dilemmas as we daily wind our way through the risky, fractious world that gave us birth. We have mixed feelings. We are not sure of this or that course of action. We understand too well that no one is so wise and great that he cannot make a catastrophic mistake, or any organization so noble to be free of corruption. We, all of us, live out our lives in conflict and contention.

The struggles born of multilevel natural selection are also where the humanities and social sciences dwell. Human beings are fascinated by other human beings, as are all other primates riveted by their own kind. We are pleased endlessly to watch and analyze our relatives, friends, and enemies. Gossip has always been the favorite occupation, in every society from hunter-gatherer bands to royal courts. To weigh as accurately as possible the intentions and trustworthiness of those who affect our own personal lives is both very human and highly adaptive. It is also adaptive to judge the impact of others' behavior on the welfare of the group as a whole. We are geniuses at reading intentions of others while they too struggle hour by hour with their own angels and demons. Civil law is the means by which we moderate the damage of our inevitable failures.

The confusion is compounded by the fact that humanity lives in a largely mythic, spirit-haunted world. We owe that to our early history. When our remote ancestors acquired a full recognition of their personal mortality, probably 100,000 to 75,000 years ago, they sought an explanation of who they were and the meaning of the world each was destined soon to leave. They must have asked, Where do the dead go? Into the spirit world, many believed. And how might we see them again? It was possible to do so at any time by dreams, or drugs, or magic, or self-inflicted privation and torture.

The early humans had no knowledge of Earth beyond the reach of their territory and trading networks. They knew nothing of the sky beyond the celestial sphere on the inner surface across which traveled the sun, moon, and stars. To explain the mysteries of their existence, they believed in the superior beings otherwise like themselves, the divine ones who built not just stone tools and shelters but the whole universe. As chiefdoms and then political states evolved, the people imagined that supernatural rulers must exist in addition to the Earth-bound rulers they followed.

The early humans needed a story of everything important that happened to them, because the conscious mind cannot work without stories and explanations of its own meaning. The best, the *only* way our forebears could manage to explain existence itself was a creation myth. And every creation myth, without exception, affirmed the superiority of the tribe that invented it over all other tribes. That much assumed, every religious believer saw himself as a chosen person.

Organized religions and their gods, although conceived in ignorance of most of the real world, were unfortunately set in stone in early history. As in the beginning, they are everywhere still an expression of tribalism by which the members establish their own identity and special relation to the supernatural world. Their dogmas codify rules of behavior that the devout can accept absolutely without hesitation. To question the sacred myths is to question the identity and worth of those who believe them. That is why skeptics, including those committed to different, equally absurd myths, are so righteously disliked. In some countries, they risk imprisonment or death.

Yet the same biological and historical circumstances that led us into the sloughs of ignorance have in other ways served humanity well. Organized religions preside over the rites of passage, from birth to maturity, from marriage to death. They offer the best a tribe has to offer: a committed community that gives heartfelt emotional support, and welcomes, and forgives. Beliefs in the gods, whether single or multiple, sacralize communal actions, including the appointment of leaders, obedience to laws, and declarations of war. Beliefs in immortality and ultimate divine justice give priceless comfort, and they steel resolution and bravery in difficult times. For millennia, organized religions have been the source of much of the best in the creative arts.

Why, then, is it wise openly to question the myths and gods of organized religions? Because they are stultifying and divisive. Because each is just one version of a competing multitude of scenarios that possibly can be true. Because they encourage ignorance, distract people from recognizing problems of the real world, and often lead them in wrong directions into disastrous actions. True to their biological origins, they passionately encourage altruism within their membership, and systematically extend it to outsiders, albeit usually with the additional aim of proselytization. Commitment to a particular faith is by definition religious bigotry. No Protestant missionary ever advises his flock to consider Roman Catholicism or Islam as a possibly superior alternative. He must by implication declare them inferior.

Yet it is foolish to think that organized religions can be pulled up anytime soon by their deep roots and replaced with a rationalist passion for morality. More likely it will happen gradually, as it is occurring in Europe, pushed along by several ongoing trends. The most potent of the trends is the increasingly detailed scientific reconstruction of religious belief as an evolutionary biological product. When placed in opposition to creation myths and their theological excesses, the reconstruction is increasingly persuasive to any even slightly open mind. Another trend against the misadventure of sectarian devotion is the growth of the internet and the globalization of institutions and people using it. A recent analysis has shown that the increasing interconnection of people worldwide strengthens their cosmopolitan attitudes. It does so by weakening the relevance of ethnicity, locality, and nationhood as sources of identification. It enhances a second trend, the homogenization of humanity in race and ethnicity through intermarriage. Inevitably, it will weaken confidence in creation myths and sectarian dogmas.

A good first step toward the liberation of humanity from the oppressive forms of tribalism would be to repudiate, respectfully, the claims of those in power who say they speak for God, are a special representative of God, or have exclusive knowledge of God's divine will. Included among these purveyors of theological narcissism are would-be prophets, the founders of religious cults, impassioned evangelical ministers, ayatollahs, imams of the grand mosques, chief rabbis, Rosh yeshivas, the Dalai Lama, and the pope. The same is true for dogmatic political ideologies based on unchallengeable precepts, left or right, and especially where justified with the dogmas of organized religions. They may contain intuitive wisdom worth hearing. Their leaders may mean well. But humanity has suffered enough from grossly inaccurate history told by mistaken prophets.

I am reminded of a story, told me long ago by a medical entomologist, about the transmission of relapsing fever by *Ornithodorus* ticks in West Africa. When the fever became severe, he said, it was the practice of the people to move the village to a new location. One day, as such an emigration was under way, he saw an elder picking up some of the ugly distant relatives of spiders off the dirt floor of a dwelling and placing them carefully in a small box. When asked why he was doing this, the man said he was transporting them to the new site, because "their spirits protect us from the fever."

Another argument for a new Enlightenment is that we are alone on this planet with whatever reason and understanding we can muster, and hence solely responsible for our actions as a species. The planet we have conquered is not just a stop along the way to a better world out there in some other dimension. Surely one moral precept we can agree on is to stop destroying our birthplace, the only home humanity will ever have. The evidence for climate warming, with industrial pollution as the principal cause, is now overwhelming. Also evident upon even casual inspection is the rapid disappearance of tropical forests and grasslands and other habitats where most of the diversity of life exists. If global changes caused by HIPPO (Habitat destruction, Invasive species, Pollution, Overpopulation, and Overharvesting, in that order of importance) are not abated, half the species of plants and animals could be extinct or at least among the "living dead"—about to become extinct—by the end of the century. We are needlessly turning the gold we inherited from our forebears into straw, and for that we will be despised by our descendants.

The obliteration of biodiversity in the living world has received much less attention than climate changes, depletion of irreplaceable resources, and other transformations of the physical environment. It would be wise to observe the following principle: if we save the living world, we will also automatically save the physical world, because in order to achieve the first we must also achieve the second. But if we save only the physical world, which appears our present inclination, we will ultimately lose them both. Until recently there existed many kinds of birds we will never again see fly. Gone are frogs we will never again hear calling on warm rainy nights. Gone are fish flashing silver in our impoverished lakes and streams.

It will be useful in taking a second look at science and religion to understand the true nature of the search for objective truth. Science is not just another enterprise like medicine or engineering or theology. It is the wellspring of all the knowledge we have of the real world that can be tested and fitted to preexisting knowledge. It is the arsenal of technologies and inferential mathematics needed to distinguish the true from the false. It formulates the principles and formulas that tie all this knowledge together. Science belongs to everybody. Its constituent parts can be challenged by anybody in the world who has sufficient information to do so. It is not just "another way of knowing" as often claimed, making it coequal with religious faith. The conflict between scientific knowledge and the teachings of organized religions is irreconcilable. The chasm will continue to widen and cause no end of trouble as long as religious leaders go on making unsupportable claims about supernatural causes of reality.

Another principle that I believe can be justified by scientific evidence so far is that nobody is going to emigrate from this planet, not ever. On a local scale—the solar system—it makes little sense to continue exploration by sending live astronauts to the moon, and much less to Mars and beyond to where simple alien life forms might reasonably be sought—on Europa, the ice-sheathed moon of Jupiter, and on fiery Enceladus, a moon of Saturn. It will be far cheaper, and entail no risk to human life, to explore space with robots. The technology is already well along, in rocket propulsion, robotics, remote analysis, and information transmission, to send robots that can do more than any human visitor, including decisions made on the spot, and to transmit images and data of the highest quality back to Earth. Granted that our spirit soars at the thought of a human being—one of *us*—walking on a celestial body like explorers on unmapped continents in times long past. Yet the real thrill will be in learning in detail what is out there, and seeing ourselves what it looks like, in crisp detail, at our virtual feet two meters away, picking up soil and possibly organisms with our virtual hands and analyzing them. We can achieve all this, and soon. To send people instead of robots would be enormously expensive, risky to human life, and inefficient—the whole of it just a circus stunt.

The same cosmic myopia exists today a fortiori in the dreams of colonizing other star systems. It is an especially dangerous delusion if we see emigration into space as a solution to be taken when we have used up this planet. It is time to ask seriously why, during the 3.5-billion-year history of the biosphere, our planet has never been visited by extraterrestrials. (Except perhaps in fuzzy UFO lights in the sky and bedroom visitors during waking nightmares.) And, why has SETI, after searching the galaxy for years, never received a message from outer space? The theoretical possibility of such a contact exists and should be continued. But imagine that on one of the billions of stars in the habitable part of the galaxy an advanced civilization arose that chose to conquer other star systems in order to expand its galactic lebensraum. That event could easily have occurred a billion years before the present. If it initiated a cycle of conquest that took a million years to reach another usable planet, and after extended exploration, another million years to send forth fleets of colonizers to several other usable planets, the ET conquering race would long ago have occupied all of the habitable segment of the galaxy, including our own solar system.

Of course, a scenario to explain the absence of extraterrestrials is that we are unique in all the galaxy going back through all those billions of years; and that we alone became capable of space travel, and so the Milky Way now awaits our conquest. That scenario is highly unlikely.

I favor another possibility. Perhaps the extraterrestrials just grew up. Perhaps they found out that the immense problems of their evolving civilizations could not be solved by competition among religious faiths, or ideologies, or warrior nations. They discovered that great problems demand great solutions, rationally achieved by cooperation among whatever factions divided them. If they accomplished that much, they would have realized that there was no need to colonize other star systems. It would be enough to settle down and explore the limitless possibilities for fulfillment on the home planet.

So, now I will confess my own blind faith. Earth, by the twenty-second century, can be turned, if we so wish, into a permanent paradise for human beings, or at least the strong beginnings of one. We will do a lot more damage to ourselves and the rest of life along the way, but out of an ethic of simple decency to one another, the unrelenting application of reason, and acceptance of what we truly are, our dreams will finally come home to stay.

AND AS FOR YOU, PAUL GAUGUIN, why did you write those lines on your painting? Of course, the ready answer I suppose is that you wanted to be very clear about the symbolization of the great range of human activity depicted in your Tahitian panorama, just in case someone might miss the point. But I sense there was something more. Perhaps you asked the three questions in such a way to imply that no answers exist, either in the civilized world you rejected and left behind or in the primitive world you adopted in order to find peace. Or again, perhaps you meant that art can go no further than what you have done; and all that was left for you to do personally was express the troubling questions in script. Let me suggest yet another reason for the mystery you left us, one not necessarily in conflict with these other conjectures. I think what you wrote is an exclamation of triumph. You had lived out your passion to travel far, to discover and embrace novel styles of visual art, to ask the questions in a new way, and from all that create an authentically original work. In this sense your career is one for the ages; it was not paid out in vain. In our own time, by bringing rational analysis and art together and joining science and humanities in partnership, we have drawn closer to the answers you sought.

D'ou Venons Nous
Que Sommes Nous
Ou Allons Nous

Acknowledgments

In the writing of this work, I have been the fortunate recipient of advice and encouragement from a great editor, Robert Weil, the years of inspired support by my agent, John Taylor Williams, and the expertise in research and manuscript preparation provided by Kathleen M. Horton.

D'où Venons Nous / Que Sommes Nous / Où Allons Nous (*Where Do We Come From? / What Are We? / Where Are We Going?*) by Paul Gauguin (1848–1903), oil on canvas, Museum of Fine Arts, Boston, Massachusetts; photograph © SuperStock.

References

1–4 PROLOGUE

Life and art of Paul Gauguin. The definitive work, by Belinda Thomson, ed., with Tamar Garb and multiple authors, is *Gauguin: Maker of Myth* (Washington, DC: Tate Publishing, National Gallery of Art, 2010).

13–20 CHAPTER 2: THE TWO PATHS TO CONQUEST

Geological origins of eusocial insect groups. Termites: Jessica L. Ware, David A. Grimaldi, and Michael S. Engel, "The effects of fossil placement and calibration on divergence times and rates: An example from the termites (Insecta: Isoptera)," *Arthropod Structure and Development* 39: 204–219 (2010). **Ants:** summary of estimates by Edward O. Wilson and Bert Hölldobler, "The rise of the ants: A phylogenetic and ecological explanation," *Proceedings of the National Academy of Sciences, U.S.A.* 102(21): 7411–7414 (2005). **Bees:** Michael Ohl and Michael S. Engel, "Die Fossilgeschichte der Bienen und ihrer nächsten Verwandten (Hymenoptera: Apoidea)," *Denisia* 20: 687–700 (2007).

Early evolution of the Old World primates. Iyad S. Zalmout et al., "New Oligocene primate from Saudi Arabia and the divergence of apes and Old World monkeys," *Nature* 466: 360–364 (2010).

21–32 CHAPTER 3: THE APPROACH

The number of individuals in the entire lineage of *Homo sapiens*. The reasoning I chose has 10^8 years as the whole geological span, and 10 years as the average longevity of a reproductive animal in the lineage to *Homo sapiens*, hence 10^7 generations in the geological span, against 10^4 individuals in each generation.

Knuckle walking versus straight-up walking. Tracy L. Kivell and Daniel Schmitt, "Independent evolution of knuckle-walking in African apes shows that humans did not evolve from a knuckle-walking ancestor," *Proceedings of the National Academy of Sciences, U.S.A.* 106(34): 14241–14246 (2009).

Persistence hunting. Louis Liebenberg, "Persistence hunting by modern hunter-gatherers," *Current Anthropology* 47(6): 1017–1025 (2006).

On persistence running by Shawn Found. Bernd Heinrich, *Racing the Antelope: What Animals Can Teach Us about Running and Life* (New York: HarperCollins, 2001).

The ability to throw objects as a preadaptation. Paul M. Bingham, "Human uniqueness: A general theory," *Quarterly Review of Biology* 74(2): 133–169 (1999).

Extinction rates in small and large mammals. Lee Hsiang Liow et al., "Higher origination and extinction rates in larger mammals," *Proceedings of the National Academy of Sciences, U.S.A.* 105(16): 6097–6102 (2008).

The fragmentation of social populations. Guy L. Bush et al., "Rapid speciation and chromosomal evolution in mammals," *Proceedings of the National Academy of Sciences, U.S.A.* 74(9): 3942–3946 (1977); Don Jay Melnick, "The genetic consequences of primate social organization," *Genetica* 73: 117–135 (1987).

33–44 CHAPTER 4: THE ARRIVAL

On *Homo habilis*. Winfried Henke, "Human biological evolution," in Franz M. Wuketits and Francisco Ayala, eds., *Handbook of Evolution*, vol. 2, *The Evolution of Living Systems* (*Including Humans*) (Weinheim: Wiley-VCH, 2005), pp. 117–222.

Climate change and early hominid evolution. Elisabeth S. Vrba et al., eds., *Paleoclimate and Evolution, with Emphasis on Human Origins* (New Haven: Yale University Press, 1995).

Chimpanzee digging tools. R. Adriana Hernandez-Aguilar, Jim Moore, and Travis Rayne Pickering, "Savanna chimpanzees use tools to harvest the underground storage organs of plants," *Proceedings of the National Academy of Sciences, U.S.A.* 104(49): 19210–19213 (2007).

Intelligence in big-brained birds. Daniel Sol et al., "Big brains, enhanced cognition, and response of birds to novel environments," *Proceedings of the National Academy of Sciences, U.S.A.* 102(15): 5460–5465 (2005).

Brain size and social organization in carnivores. John A. Finarelli and John J. Flynn, "Brain-size evolution and sociality in Carnivora," *Proceedings of the National Academy of Sciences, U.S.A.* 106(23): 9345–9349 (2009).

Ancient tools. J. Shreeve, "Evolutionary road," *National Geographic* 218: 34–67 (July 2010).

The evolutionary shift to meat eating. David R. Braun et al., "Early hominin diet included diverse terrestrial and aquatic animals 1.95

Ma in East Turkana, Kenya," *Proceedings of the National Academy of Sciences, U.S.A.* 107(22): 10002–10007 (2010); Teresa E. Steele, "A unique hominin menu dated to 1.95 million years ago," *Proceedings of the National Academy of Sciences, U.S.A.* 107(24): 10771–10772 (2010).

Bonobo predation. Martin Surbeck and Gottfried Hohmann, "Primate hunting by bonobos at LuiKotale, Salonga National Park," *Current Biology* 18(19): R906–R907 (2008).

Neanderthals as big-game hunters. Michael P. Richards and Erik Trinkaus, "Isotopic evidence for the diets of European Neanderthals and early modern humans," *Proceedings of the National Academy of Sciences, U.S.A.* 106(38): 16034–16039 (2009). Neanderthals also consumed a variety of vegetable foods when available: Amanda G. Henry, Alison S. Brooks, and Dolores R. Piperno, "Microfossils in calculus demonstrate consumption of plants and cooked foods in Neanderthal diets (Shanidar III, Iraq; Spy I and II, Belgium)," *Proceedings of the National Academy of Sciences, U.S.A.* 108(2): 486–491 (2011).

49–56 CHAPTER 6: THE CREATIVE FORCES

Kin selection in human evolution. In the 1970s, I was one of the scientists promoting kin selection as central in the origin of eusociality and human evolution, in *Sociobiology: The New Synthesis* (Cambridge, MA: Belknap Press of Harvard University Press, 1975) and *On Human Nature* (Cambridge, MA: Harvard University Press, 1978). To the extent I emphasized it, I now believe, I was wrong. See Edward O. Wilson, "One giant leap: How insects achieved altruism and colonial life," *BioScience* 58(1): 17–25 (2008); Martin A. Nowak, Corina E. Tarnita, and Edward O. Wilson, "The evolution of eusociality," *Nature* 466: 1057–1062 (2010).

A new theory of eusocial evolution, including queen-to-queen selection in the social insects. Martin A. Nowak, Corina E. Tarnita, and Edward O. Wilson, "The evolution of eusociality," *Nature* 466: 1057–1062 (2010).

57–61 CHAPTER 7: TRIBALISM IS A FUNDAMENTAL HUMAN TRAIT

Exuberance over an athletic victory. Roger Brown, *Social Psychology* (New York: Free Press, 1965; 2nd ed. 1985), p. 553.

In-group formation as instinct. Roger Brown, *Social Psychology* (New York: Free Press, 1965; 2nd ed. 1985), p. 553; Edward O. Wilson, *Consilience: The Unity of Knowledge* (New York: Knopf, 1998).

The preference for native language in the formation of groups.

Katherine D. Kinzler, Emmanuel Dupoux, and Elizabeth S. Spelke, "The native language of social cognition," *Proceedings of the National Academy of Sciences, U.S.A.* 104(30): 12577–12580 (2007).

Brain activation and control of fear. Jeffrey Kluger, "Race and the brain," *Time*, p. 59 (20 October 2008).

62–76 CHAPTER 8: WAR AS HUMANITY'S HEREDITARY CURSE

William James on war. William James, "The moral equivalent of war," *Popular Science Monthly* 77: 400–410 (1910).

War and genocide by the U.S.S.R. and Nazi Germany. Timothy Snyder, "Holocaust: The ignored reality," *New York Review of Books* 56(12) (16 July 2009).

Martin Luther on God's use of war. Martin Luther in *Whether Soldiers, Too, Can Be Saved* (1526), trans. J. M. Porter, *Luther: Selected Political Writings* (Lanham, MD: University Press of America, 1988), p. 103.

Athenians conquer Melos. William James, "The moral equivalent of war," *Popular Science Monthly* 77: 400–410 (1910); Thucydides, *The Peloponnesian War*, trans. Walter Banco (New York: W. W. Norton, 1998). The wording quoted here is from the translation used by William James.

Evidences of prehistoric warfare. Steven A. LeBlanc and Katherine E. Register, *Constant Battles: The Myth of the Peaceful, Noble Savage* (New York: St. Martin's Press, 2003).

Buddhism and war. Bernard Faure, "Buddhism and violence," *International Review of Culture & Society* no. 9 (Spring 2002); Michael Zimmermann, ed., *Buddhism and Violence* (Bhairahana, Nepal: Lumbini International Research Institute, 2006).

Persistence of war. Steven A. LeBlanc and Katherine E. Register, *Constant Battles: The Myth of the Peaceful, Noble Savage* (New York: St. Martin's Press, 2003).

Early models of group selection. Richard Levins, "The theory of fitness in a heterogeneous environment, IV: The adaptive significance of gene flow," *Evolution* 18(4): 635–638 (1965); Richard Levins, *Evolution in Changing Environments: Some Theoretical Explorations* (Princeton, NJ: Princeton University Press, 1968); Scott A. Boorman and Paul R. Levitt, "Group selection on the boundary of a stable population," *Theoretical Population Biology* 4(1): 85–128 (1973); Scott A. Boorman and P. R. Levitt, "A frequency-dependent natural selection model for the evolution of social cooperation networks," *Proceedings of the National Academy of Sciences, U.S.A.* 70(1): 187–189 (1973). The foregoing articles were reviewed by Edward O. Wilson, *Sociobiology: The New Synthesis* (Cambridge, MA: Belknap Press of Harvard University Press, 1975), pp. 110–117.

Violence and death in humans and chimpanzees. Richard W. Wrangham, Michael L. Wilson, and Martin N. Muller, "Comparative rates of violence in chimpanzees and humans," *Primates* 47: 14–26 (2006).

Comparison of human and chimpanzee aggression. Richard W. Wrangham and Michael L. Wilson, "Collective violence: Comparison between youths and chimpanzees," *Annals of the New York Academy of Science* 1036: 233–256 (2004).

Chimpanzee warfare. John C. Mitani, David P. Watts, and Sylvia J. Amsler, "Lethal intergroup aggression leads to territorial expansion in wild chimpanzees," *Current Biology* 20(12): R507–R508 (2010). An excellent report and commentary is provided by Nicholas Wade in "Chimps that wage war and annex rival territory," *New York Times*, D4 (22 June 2010).

Population control. The concept of the *minimum-limiting factor* was introduced by Carl Sprengel in 1828 for agriculture, and later formalized by Justus von Liebig—hence it is sometimes called Liebig's law of the minimum. In the original formulation, it said that the growth of crops is determined not by the total amount of nutrients but by the scarcest one among them.

Demographic shocks and alliance formation. E. A. Hammel, "Demographics and kinship in anthropological populations," *Proceedings of the National Academy of Sciences, U.S.A.* 102(6): 2248–2253 (2005).

Human population size, regional limits. R. Hopfenberg, "Human carrying capacity is determined by food availability," *Population and Environment* 25: 109–117 (2003).

77–84 CHAPTER 9: THE BREAKOUT

***Home erectus* footprints**. Report in "World Roundup: Archaeological assemblages: Kenya," *Archaeology*, p. 11 (May/June 2009).

Appearance of modern *Homo sapiens*. G. Philip Rightmire, "Middle and later Pleistocene hominins in Africa and Southwest Asia," *Proceedings of the National Academy of Sciences, U.S.A.* 106(38): 16046–16050 (2009).

Genomes of Africans. Stephan C. Schuster et al., "Complete Kohisan and Bantu genomes from southern Africa," *Nature* 463: 943–947 (2010).

85–96 CHAPTER 10: THE CREATIVE EXPLOSION

Serial founder effect in human emigration. Sohini Ramachandran et al., "Support from the relationship of genetic and geographic distance in human populations for a serial founder effect originating in Africa," *Proceedings of the National Academy of Sciences, U.S.A.* 102(44): 15942–15947 (2005).

Genetic sweep of emigrants up the Nile. Henry Harpending and Alan Rogers, "Genetic perspectives on human origins and differentiation," *Annual Review of Genomics and Human Genetics* 1: 361–385 (2000).

Climate changes and the out-of-Africa spread. Andrew S. Cohen et al., "Ecological consequences of early Late Pleistocene megadroughts in tropical Africa," *Proceedings of the National Academy of Sciences, U.S.A.* 104(42): 16428–16427 (2007).

***Homo sapiens* enter Europe and Neanderthals disappear**. John F. Hoffecker, "The spread of modern humans in Europe," *Proceedings of the National Academy of Sciences, U.S.A.* 106(38): 16040–16045 (2009); J. J. Hublin, "The origin of Neandertals," *Proceedings of the National Academy of Sciences, U.S.A.* 106(38): 16022–16027 (2009).

Discovery of a new hominin, the "Denisovans." David Reich et al., "Genetic history of an archaic hominin group from Denisova Cave in Siberia," *Nature* 468: 1053–1060 (2010).

Spread of *Homo sapiens* in the Old World. Peter Foster and S. Matsumura, "Did early humans go north or south?" *Science* 308: 965–966 (2005); Cristopher N. Johnson, "The remaking of Australia's ecology," *Science* 309: 255–256; Gifford H. Miller et al., "Ecosystem collapse in Pleistocene Australia and a human role in megafaunal extinction," *Science* 309: 287–290 (2005).

Human invasion of the New World. Ted Goebel, Michael R. Waters, and Dennis H. O'Rourke, "The Late Pleistocene dispersal of modern humans in the Americas," *Science* 319: 1497–1502 (2008); Andrew Curry, "Ancient excrement," *Archaeology*, pp. 42–45 (July/August 2008).

Discontinuities in cultural innovation. Francesco d'Errico et al., "Additional evidence on the use of personal ornaments in the Middle Paleolithic of North Africa," *Proceedings of the National Academy of Sciences, U.S.A.* 106(38): 16051–16056 (2009).

Rate of evolution increases with humanity's spread. John Hawks et al., "Recent acceleration of human adaptive evolution," *Proceedings of the National Academy of Sciences, U.S.A.* 104(52): 20753–20758 (2007).

Adaptive evolution in recent human evolution. Jun Gojobori et al., "Adaptive evolution in humans revealed by the negative correlation between the polymorphism and fixation phases of evolution," *Proceedings of the National Academy of Sciences, U.S.A.* 104(10): 3907–3912 (2007).

Changes in frequency of mutant genes. Jun Gojobori et al., "Adaptive evolution in humans revealed by the negative correlation between the polymorphism and fixation phases of evolution," *Proceedings of the National Academy of Sciences, U.S.A.* 104(10): 3907–3912 (2007).

Genes in the evolution of human cognition. Ralph Haygood et al., "Contrasts between adaptive coding and noncoding changes during human evolution," *Proceedings of the National Academy of Sciences, U.S.A.* 107(17): 7853–7857 (2010).

Genetic inheritance of mental traits. B. Devlin, Michael Daniels, and Kathryn Roeder, "The heritability of IQ," *Nature* 388: 468–471 (1997). Various estimates for IQ fall between 0.4 and 0.7, most likely toward the lower value.

Turkheimer's first law. E. Turkheimer, "Three laws of behavior genetics and what they mean," *Current Directions in Psychological Science* 9(5): 160–164 (2000).

Genetic factors in networking. James Fowler, Christopher T. Dawes, and Nicholas A. Christakis, "Model of genetic variation in human social networks," *Proceedings of the National Academy of Sciences, U.S.A.* 106(6): 1720–1724 (2009).

Concepts invented in or before the Neolithic period. Dwight Read and Sander van der Leeuw, "Biology is only part of the story," *Philosophical Transactions of the Royal Society* **B** 363: 1959–1968 (2008).

Origin of domestic plants. Colin E. Hughes et al., "Serendipitous backyard hybridization and the origin of crops," *Proceedings of the National Academy of Sciences, U.S.A.* 104(36): 14389–14394 (2007).

Natural selection in contemporary humans. Steve Olson, "Seeking the signs of selection," *Science* 298: 1324–1325 (2002); Michael Balter, "Are humans still evolving?" *Science* 309: 234–237 (2005); Cynthia M. Beall et al., "Natural selection on *EPAS1 (H1F2α)* associated with low hemoglobin concentration in Tibetan highlanders," *Proceedings of the National Academy of Sciences, U.S.A.* 107(25): 11459–11464 (2010); Oksana Hlodan, "Evolution in extreme environments," *BioScience* 60(6): 414–418 (2010).

97–105 CHAPTER 11: THE SPRINT TO CIVILIZATION

The sprint to civilization from bands to states. Kent V. Flannery, "The cultural evolution of civilizations," *Annual Review of Ecology and*

Systematics 3: 399–426 (1972); H. T. Wright, "Recent research on the origin of the state," *Annual Review of Anthropology* 6: 379–397 (1977); Charles S. Spencer, "Territorial expression and primary state formation," *Proceedings of the National Academy of Sciences, U.S.A.* 107: 7119–7126 (2010).

Simon's principle of hierarchies. Herbert A. Simon, "The architecture of complexity," *Proceedings of the American Philosophical Society* 106: 467–482 (1962).

Variation of personality in Burkina Faso. Richard W. Robins, "The nature of personality: genes, culture, and national character," *Science* 310: 62–63 (2005).

Personality variation within and across cultures. A. Terraciano et al., "National character does not reflect mean personality trait levels in 49 cultures," *Science* 310: 96–100 (2005).

Origin times of state-based civilization. Charles S. Spencer, "Territorial expansion and primary state formation," *Proceedings of the National Academy of Sciences, U.S.A.* 107(16): 7119–7126 (2010).

Dates of the origin of primary states. Charles S. Spencer, "Territorial expansion and primary state formation," *Proceedings of the National Academy of Sciences, U.S.A.* 107(16): 7119–7126 (2010).

Swift origin of a primary state in Hawaii. Patrick V. Kirch and Warren D. Sharp,"Coral ^{230}Th dating of the imposition of a ritual control hierarchy in precontact Hawaii," *Science* 307: 102–104 (2005).

Engraved eggshell containers. Pierre-Jean Texier et al., "A Howiesons Poort tradition of engraving ostrich eggshell containers dated to 60,000 years ago at Diepkloof Rock Shelter, South Africa," *Proceedings of the National Academy of Sciences, U.S.A.* 107(14): 6180–6185 (2010).

Earliest African art and weaponry. Constance Holden, "Oldest beads suggest early symbolic behavior," *Science* 304: 369 (2004); Christopher Henshilwood et al., "Middle Stone Age shell beads from South Africa," *Science* 304: 404 (2004).

Ancient temple at Göbekli Tepe. Andrew Curry, "Seeking the roots of ritual," *Science* 319: 278–280 (2008).

Origin of writing. Andrew Lawler, "Writing gets a rewrite," *Science* 292: 2418–2420 (2001); John Noble Wilford, "Stone said to contain earliest writing in Western Hemisphere," *New York Times*, A12 (15 September 2006).

Meaning of ancient script. Barry B. Powell, *Writing: Theory and History of the Technology of Civilization* (Malden, MA: Wiley-Blackwell, 2009).

Cultural evolution and the origin of the Neolithic period. Jared Diamond, *Guns, Germs, and Steel: The Fates of Human Societies* (New York: W. W. Norton, 1997); Douglas A. Hibbs Jr. and Ola Olsson, "Geography, biogeography, and why some countries are rich and others are poor," *Proceedings of the National Academy of Sciences, U.S.A.* 101(10): 3715–3720 (2004).

109–119 CHAPTER 12: THE INVENTION OF EUSOCIALITY

Social insect dominance of Amazonian Forest. H. J. Fittkau and H. Klinge, "On biomass and trophic structure of the central Amazonian rainforest ecosystem," *Biotropica* 5: 2–14 (1973).

120–130 CHAPTER 13: INVENTIONS THAT ADVANCED THE SOCIAL INSECTS

Migratory ants and herds of sapsuckers. U. Maschwitz, M. D. Dill, and J. Williams, "Herdsmen ants and their mealybug partners," *Abhandlungen der Senckenbergischen Naturforschenden Gesellschaft Frankfurt am Main* 557: 1–373 (2002).

133–138 CHAPTER 16: THE SCIENTIFIC DILEMMA OF RARITY

The evolutionary origin of eusociality. Edward O. Wilson and Bert Hölldobler, "Eusociality: Origin and consequences," *Proceedings of the National Academy of Sciences, U.S.A.* 102(38): 13367–13371 (2005); Charles D. Michener, *The Bees of the World* (Baltimore: Johns Hopkins University Press, 2007); Bryan N. Danforth, "Evolution of sociality in a primitively eusocial lineage of bees," *Proceedings of the National Academy of Sciences, U.S.A.* 99(1): 286–290 (2002); Bert Hölldobler and Edward O. Wilson, *The Superorganism: The Beauty, Elegance, and Strangeness of Insect Societies* (New York: W. W. Norton, 2009).

Eusociality in shrimps. J. Emmett Duffy, C. L. Morrison, and R. Ríos, "Multiple origins of eusociality among sponge-dwelling shrimps (*Synalpheus*)," *Evolution* 54(2): 503–516 (2000).

Unique evolutionary events. Geerat J. Vermeij, "Historical contingency and the purported uniqueness of evolutionary innovations," *Proceedings of the National Academy of Sciences, U.S.A.* 103(6): 1804–1809 (2006).

Helpers at the nest in birds. B. J. Hatchwell and J. Komdeur, "Ecological constraints, life history traits and the evolution of cooperative breeding," *Animal Behaviour* 59(6): 1079–1086 (2000).

139–147 CHAPTER 15: INSECT ALTRUISM AND EUSOCIALITY EXPLAINED

Origins of insect societies. William Morton Wheeler, *Colony Founding among Ants, with an Account of Some Primitive Australian Species* (Cambridge, MA: Harvard University Press, 1933); Charles D. Michener, "The evolution of social behavior in bees," *Proceedings of the Tenth International Congress in Entomology, Montreal* 2: 441–447 (1956); Howard E. Evans, "The evolution of social life in wasps," *Proceedings of the Tenth International Congress in Entomology, Montreal,* 2: 449–457 (1956).

Replacing kin selection. Martin A. Nowak, Corina E. Tarnita, and Edward O. Wilson, "The evolution of eusociality," *Nature* 466: 1057–1062 (2010). A later account is provided by Martin A. Nowak and Roger Highfield in *SuperCooperators: Altruisim, Evolution, and Why We Need Each Other to Succeed* (New York: Free Press, 2011).

The steps to eusociality in the insects. Edward O. Wilson, "One giant leap: How insects achieved altruism and colonial life," *BioScience* 58: 17–25 (2008).

Natural resources and early eusociality in insects. Edward O. Wilson and Bert Hölldobler, "Eusociality: Origin and consequences," *Proceedings of the National Academy of Sciences, U.S.A.* 102(38): 13367–13371 (2005).

Solitary Hymenoptera. James T. Costa, *The Other Insect Societies* (Cambridge, MA: Belknap Press of Harvard University Press, 2006).

Eusocial beetles. D. S. Kent and J. A. Simpson, "Eusociality in the beetle *Austroplatypus incompertus* (Coleoptera: Curculionidae)," *Naturwissenschaften* 79: 86–87 (1992).

Eusocial thrips and aphids. Bernard J. Crespi, "Eusociality in Australian gall thrips," *Nature* 359: 724–726 (1992); David L. Stern and W. A. Foster, "The evolution of soldiers in aphids," *Biological Reviews of the Cambridge Philosophical Society* 71: 27–79 (1996).

Eusocial shrimp. J. Emmett Duffy, "Ecology and evolution of eusociality in sponge-dwelling shrimp," in J. Emmett Duffy and Martin Thiel, eds., *Evolutionary Ecology of Social and Sexual Systems: Crustaceans as Model Organisms* (New York: Oxford University Press, 2007).

Artifically induced eusocial bee colonies. Shoichi F. Sakagami and Yasuo Maeta, "Sociality, induced and/or natural, in the basically solitary small carpenter bees (*Ceratina*)," in Yosiaki Itô, Jerram L. Brown, and Jiro Kikkawa, eds., *Animal Societies: Theories and Facts* (Tokyo: Japan Scientific Societies Press, 1987), pp. 1–16; William

T. Wcislo, "Social interactions and behavioral context in a largely solitary bee, *Lasioglossum (Dialictus) figueresi* (Hymenoptera, Halictidae)," *Insectes Sociaux* 44: 199–208 (1997); Raphael Jeanson, Penny F. Kukuk, and Jennifer H. Fewell, "Emergence of division of labour in halictine bees: Contributions of social interactions and behavioural variance," *Animal Behaviour* 70: 1183–1193 (2005).

Fixed-threshold model in division of labor, insects. Gene E. Robinson and Robert E. Page Jr., "Genetic basis for division of labor in an insect society," in Michael D. Breed and Robert E. Page Jr., eds., *The Genetics of Social Evolution* (Boulder, CO: Westview Press 1989), pp. 61–80; E. Bonabeau, G. Theraulaz, and Jean-Luc Deneubourg, "Quantitative study of the fixed threshold model for the regulation of division of labour in insect societies," *Proceedings of the Royal Society* **B** 263: 1565–1569 (1996); Samuel N. Beshers and Jennifer H. Fewell, "Models of division of labor in social insects," *Annual Review of Entomology* 46: 413–440 (2001).

148–157 CHAPTER 16: INSECTS TAKE THE GIANT LEAP

The value of nest defense. J. Field and S. Brace, "Pre-social benefits of extended parental care," *Nature* 427: 650–652 (2004).

Back-and-forth social evolution in bees. Bryan N. Danforth, "Evolution of sociality in a primitively eusocial lineage of bees," *Proceedings of the National Academy of Sciences, U.S.A.* 99(1): 286–290 (2002).

Seasonal change promotes social behavior. James H. Hunt and Gro V. Amdam, "Bivoltinism as an antecedent to eusociality in the paper wasp genus *Polistes*," *Science* 308: 264–267 (2005).

The origin of wingless workers in the ants. Ehab Abouheif and G. A. Wray, "Evolution of the gene network underlying wing polyphenism in ants," *Science* 297: 249–252 (2002).

The origin of polygyny in fire ants. Kenneth G. Ross and Laurent Keller, "Genetic control of social organization in an ant," *Proceedings of the National Academy of Sciences, U.S.A.* 95(24): 14232–14237 (1998).

Genes and eusocial behavior in fire ants. M. J. B. Krieger and Kenneth G. Ross, "Identification of a major gene regulating complex social behavior," *Science* 295: 328–332 (2002).

Genetics and development in social wasps. James H. Hunt and Gro V. Amdam, "Bivoltinism as an antecedent to eusociality in the paper wasp genus *Polistes*," *Science* 308: 264–267 (2005).

Cooperation among solitary bees. Shoichi F. Sakagami and Yasuo Maeta, "Sociality, induced and/or natural, in the basically solitary small carpenter bees (*Ceratina*)," in Yosiaki Itô, Jerram L. Brown, and Jiro Kikkawa, eds., *Animal Societies: Theories and Facts* (Tokyo: Japan Scientific Societies Press, 1987), pp. 1–16.

Cooperating queens in primitively eusocial bees. Miriam H. Richards, Eric J. von Wettberg, and Amy C. Rutgers, "A novel social polymorphism in a primitively eusocial bee," *Proceedings of the National Academy of Sciences, U.S.A.* 100(12): 7175–7180 (2003).

Reversal of the sequence in the ground plan leads to eusociality. Gro V. Amdam et al., "Complex social behaviour from maternal reproductive traits," *Nature* 439: 76–78 (2006); Gro V. Amdam et al., "Variation in endocrine signaling underlies variation in social life-history," *American Naturalist* 170: 37–46 (2007).

The point of no return in eusocial evolution. Edward O. Wilson, *The Insect Societies* (Cambridge, MA: Belknap Press of Harvard University Press, 1971); Edward O. Wilson and Bert Hölldobler, "Eusociality: Origin and consequence," *Proceedings of the National Academy of Sciences, U.S.A.* 102(38): 13367–13371 (2005).

158–165 CHAPTER 17: HOW NATURAL SELECTION CREATES SOCIAL INSTINCTS

Darwin on instincts as genetic adaptations. Darwin's great books: in addition to *The Expression of the Emotions in Man and Animals* (1873), the other three were *Voyage of the Beagle* (1838), *Origin of Species* (1859), and *The Descent of Man* (1872).

166–182 CHAPTER 18: THE FORCES OF SOCIAL EVOLUTION

Hamilton on kin selection. William D. Hamilton, "The genetical evolution of social behaviour, I, II," *Journal of Theoretical Biology* 7: 1–52 (1964).

Haldane's formulation of kin selection. J. B. S. Haldane, "Population genetics," *New Biology* (Penguin Books) 18: 34–51 (1955).

The failure of the haplodiploid hypothesis. Edward O. Wilson, "One giant leap: How insects achieved altruism and colonial life," *BioScience* 58(1): 17–25 (2008).

Advantages of genetic diversity in ant colonies. Blaine Cole and Diane C. Wiernacz, "The selective advantage of low relatedness," *Science* 285: 891–893 (1999); William O. H. Hughes and J. J. Boomsma, "Genetic diversity and disease resistance in leaf-cutting ant societies," *Evolution* 58: 1251–1260 (2004).

Genetically diverse ant castes. F. E. Rheindt, C. P. Strehl, and Jürgen Gadau, "A genetic component in the determination of worker polymorphism in the Florida harvester ant *Pogonomyrmex badius*," *Insectes Sociaux* 52: 163–168 (2005).

Climate control in social insect nests. J. C. Jones, M. R. Myerscough, S. Graham, and Ben P. Oldroyd, "Honey bee nest thermoregulation: Diversity supports stability," *Science* 305: 402–404 (2004).

Genetic factors in division of labor within ant colonies. T. Schwander, H. Rosset, and M. Chapuisat, "Division of labour and worker size polymorphism in ant colonies: The impact of social and genetic factors," *Behavioral Ecology and Sociobiology* 59: 215–221 (2005).

Sequenced multilevel theory owes its origin to many sources, but the main thrust of its development occurred through the following articles, in which the present author played a role. Edward O. Wilson, "Kin selection as the key to altruism: Its rise and fall," *Social Research* 72(1): 159–166 (2005); Edward O. Wilson and Bert Hölldobler, "Eusociality: Origin and consequences," *Proceedings of the National Academy of Sciences, U.S.A.* 102(38): 13367–13371 (2005); David Sloan Wilson and Edward O. Wilson, "Rethinking the theoretical foundation of sociobiology," *Quarterly Review of Biology* 82(4): 327–348 (2007); Edward O. Wilson, "One giant leap: How insects achieved altruism and colonial life," *BioScience* 58(1): 17–25 (2008); David Sloan Wilson and Edward O. Wilson, "Evolution 'for the good of the group,' " *American Scientist* 96: 380–389 (2008); and finally and definitively, Martin A. Nowak, Corina E. Tarnita and Edward O. Wilson, "The evolution of eusociality," *Nature* 466: 1057–1062 (2010). The current text has drawn heavily on the last of these articles.

Sex ratio investments in social insects. Robert L. Trivers and Hope Hare, "Haplodiploidy and the evolution of the social insects," *Science* 191: 249–263 (1976); Andrew F. G. Bourke and Nigel R. Franks, *Social Evolution in Ants* (Princeton, NJ: Princeton University Press, 1995).

Dominance behavior and policing in social insects. Francis L. W. Ratnieks, Kevin R. Foster, and Tom Wenseleers, "Conflict resolution in insect societies," *Annual Review of Entomology* 51: 581–608 (2006).

Numbers of matings by each queen of social insects. William O. H. Hughes et al., "Ancestral monogamy shows kin selection is key to the evolution of eusociality," *Science* 320: 1213–1216 (2008).

Contributions of inclusive-fitness theory. Edward O. Wilson, "One giant leap: How insects achieved altruism and colonial life," *BioScience* 58: 17–25 (2008); Bert Hölldobler and Edward O. Wilson, *The*

Superorganism: The Beauty, Elegance, and Strangeness of Insect Societies (New York: W. W. Norton, 2009).

The concept of kinship used by inclusive-fitness theory. This account and much of the remainder of the chapter is modified from Martin A. Nowak, Corina E. Tarnita, and Edward O. Wilson, "The evolution of eusociality," *Nature* 466: 1057–1062 (2010).

Varying definitions of kinship. Raghavendra Gadagkar, *The Social Biology of* Ropalidia marginata: *Toward Understanding the Evolution of Eusociality* (Cambridge, MA: Harvard University Press, 2001); Barbara L. Thorne, Nancy L. Breisch, and Mario L. Muscedere, "Evolution of eusociality and the soldier caste in termites: Influence of accelerated inheritance," *Proceedings of the National Academy of Sciences, U.S.A.* 100: 12808–12813 (2003); Abderrahman Khila and Ehab Abouheif, "Evaluating the role of reproductive constraints in ant social evolution," *Philosophical Transactions of the Royal Society* **B** 365: 617–630 (2010).

The failure of Hamilton's inequality in social theory. Arne Traulsen, "Mathematics of kin- and group-selection: Formally equivalent?," *Evolution* 64: 316–323 (2010).

Critique of inclusive-fitness theory. Martin A. Nowak, Corina E. Tarnita, and Edward O. Wilson, "The evolution of eusociality," *Nature* 466: 1057–1062 (2010). See also Martin A. Nowak and Roger Highfield, *SuperCooperators: Altruism, Evolution, and Why We Need Each Other to Succeed* (New York: Free Press, 2011).

Weak selection in social evolution. Martin A. Nowak, Corina E. Tarnita, and Edward O. Wilson, "The evolution of eusociality," *Nature* 466: 1057–1062 (2010).

Alternative theories of social evolution. Martin A. Nowak, Corina E. Tarnita, and Edward O. Wilson, "The evolution of eusociality," *Nature* 466: 1057–1062 (2010).

Group selection in microorganisms. The driving force of evolution in eusocial microorganisms. A review of literature and presentation of opposing theories is provided by David Sloan Wilson and Edward O. Wilson, "Rethinking the theoretical foundations of sociobiology," *Quarterly Review of Biology* 82(4): 327–348 (2007).

Monogamy and kin selection. W. O. H. Hughes et al., "Ancestral monogamy shows kin selection is key to the evolution of eusociality," *Science* 320: 1213–1216 (2008).

Multiple matings and large colonies in the social insects. Bert Hölldobler and Edward O. Wilson, *The Superorganism: The Beauty, Elegance, and Strangeness of Insect Societies* (New York: W. W. Norton, 2009).

Kin selection postulated for policing in social insects. Francis L. W. Ratnieks, Kevin R. Foster, and Tom Wenseleers, "Conflict resolution in insect societies," *Annual Review of Entomology* 51: 581–608 (2006).

Sex investment ratios in social insects proposed. Robert L. Trivers and Hope Hare, "Haplodiploidy and the evolution of the social insects," *Science* 191: 249–263 (1976).

Sex investment ratio analyzed. Andrew F. G. Bourke and Nigel R. Franks, *Social Evolution in Ants* (Princeton, NJ: Princeton University Press, 1995).

Subsocial spiders. J. M. Schneider and T. Bilde, "Benefits of cooperation with genetic kin in a subsocial spider," *Proceedings of the National Academy of Sciences, U.S.A.* 105(31): 10843–10846 (2008).

Helpers at the nest: birds. Stuart A. West, A. S. Griffin, and A. Gardner, "Evolutionary explanations for cooperation," *Current Biology* 17: R661–R672 (2007).

Natural history, bird helpers. B. J. Hatchwell and J. Komdeur, "Ecological constraints, life history traits and the evolution of cooperative breeding," *Animal Behaviour* 59(6): 1079–1086 (2000).

183–187 CHAPTER 19: ITHE EMERGENCE OF A NEW THEORY OF EUSOCIALITY

The formation of elementary social groups. J. W. Pepper and Barbara Smuts, "A mechanism for the evolution of altruism among nonkin: Positive assortment through environmental feedback," *American Naturalist* 160: 205–213 (2002); J. A. Fletcher and M. Zwick, "Strong altruism can evolve in randomly formed groups," *Journal of Theoretical Biology* 228: 303–313 (2004).

Primitive termite social organization. Barbara L. Thorne, Nancy L. Breisch, and Mario L. Muscedere, "Evolution of eusociality and the soldier caste in termites: Influence of accelerated inheritance," *Proceedings of the National Academy of Sciences, U.S.A.* 100: 12808–12813 (2003).

Worker ants as robots. Martin A. Nowak, Corina E. Tarnita, and Edward O. Wilson, "The evolution of eusociality," *Nature* 466: 1057–1062 (2010).

Group selection and the superorganism. Bert Hölldobler and Edward O. Wilson, *The Superorganism: The Beauty, Elegance, and Strangeness of Insect Societies* (New York: W. W. Norton, 2009).

191–211 CHAPTER 20: WHAT IS HUMAN NATURE?

Introduction of the theory of gene-culture coevolution. Charles J. Lumsden and Edward O. Wilson, "Translation of epigenetic rules of individual behavior into ethnographic patterns," *Proceedings of the National Academy of Sciences, U.S.A.* 77(7): 4382–4386 (1980); "Gene-culture translation in the avoidance of sibling incest," *Proceedings of the National Academy of Sciences, U.S.A.* 77(10): 6248–6250 (1980); *Genes, Mind, and Culture: The Coevolutionary Process* (Cambridge, MA: Harvard University Press, 1981); Edward O. Wilson, *Biophilia* (Cambridge, MA: Harvard University Press, 1984).

Extensions of gene-culture theory. Charles J. Lumsden and Edward O. Wilson, *Promethean Fire: Reflection on the Origin of the Mind* (Cambridge, MA: Harvard University Press, 1983).

Genes and culture. Luigi Luca Cavalli-Sforza and Marcus W. Feldman, *Cultural Transmission and Evolution: A Quantitative Approach* (Princeton, NJ: Princeton University Press, 1981); Robert Boyd and Peter J. Richerson, *Culture and the Evolutionary Process* (Chicago: University of Chicago Press, 1985). In 1976, Marcus W. Feldman and Luigi L. Cavalli-Sforza published an analysis, "Cultural and biological evolutionary processes, selection for a trait under complex transmission," *Theoretical Population Biology* 9: 238–259 (1976), and "The evolution of continuous variation, II: Complex transmission and assortative mating," *Theoretical Population Biology* 11: 161–181 (1977), in which two states occur, "skilled" and "unskilled," the probabilities of which depend on the parent's phenotype and the offspring's genotype. The trait is one of general capability. Then, as later, no attention was paid to the abundance of data on the epigenetic rules embedded in human cognition. The history of this and other earlier work relevant to gene-culture coevolution is summarized in Charles J. Lumsden and Edward O. Wilson, *Genes, Mind, and Culture: The Coevolutionary Process* (Cambridge, MA: Harvard University Press, 1981), pp. 258–263.

Evolution of adult lactose tolerance. Sarah A. Tishkoff et al., "Convergent adaptation of human lactase persistence in Africa and Europe," *Nature Genetics* 39(1): 31–40 (2007).

Gene-culture coevolution and the expansions of diet. Olli Arjama and Tima Vuoriselo, "Gene-culture coevolution and human diet," *American Scientist* 98: 140–146 (2010).

The evolution of human diet. Richard Wrangham, *Catching Fire: How Cooking Made Us Human* (New York: Basic Books, 2009).

Gene-culture coevolution and the avoidance of incest. The account of incest avoidance given here is drawn mostly from Edward O. Wilson, *Consilience: The Unity of Knowledge* (New York: Knopf, 1998), updated with recent literature.

Evidence for Westermarck effect. Arthur P. Wolf, *Sexual Attraction and Childhood Association: A Chinese Brief for Edward Westermarck* (Stanford, CA; Stanford University Press, 1995); Joseph Shepher, "Mate selection among second generation kibbutz adolescents and adults: Incest avoidance and negative imprinting," *Archives of Sexual Behavior* 1(4): 293–307 (1971); William H. Durham, *Coevolution: Genes, Culture, and Human Diversity* (Stanford, CA: Stanford University Press, 1991).

Diseases caused by inbreeding. Jennifer Couzain and Joselyn Kaiser, "Closing the net on common disease genes," *Science* 316: 820–822 (2007); Ken N. Paige, "The functional genomics of inbreeding depression: A new approach to an old problem," *BioScience* 60: 267–277 (2010).

Exogamy and the Westermarck effect. The many cultural implications of human exogamy born from incest avoidance are the subject of a treatise by Bernard Chapais, *Primeval Kinship: How Pair-Bonding Gave Rise to Human Society* (Cambridge, MA: Harvard University Press, 2008).

An alternative explanation to the Westermarck effect. William H. Durham, *Coevolution: Genes, Culture, and Human Diversity* (Stanford, CA: Stanford University Press, 1991).

Definition of "epigenetic" and "epigenetic rules." Charles J. Lumsden and Edward O. Wilson, *Genes, Mind, and Culture: The Coevolutionary Process* (Cambridge, MA: Harvard University Press, 1981); Tabitha M. Powledge, "Epigenetics and development," *BioScience* 59: 736–741 (2009).

Color vision. The account of color vision and vocabulary here is drawn largely from Edward O. Wilson, *Consilience: The Unity of Knowledge* (New York: Knopf, 1998), updated and with additional references.

Cross-cultural color classification. Brent Berlin and Paul Kay, *Basic Color Terms: Their Universality and Evolution* (Berkeley: University of California Press, 1969).

New Guinea experiment on color classification. Eleanor Rosch, Carolyn Mervis, and Wayne Gray, *Basic Objects in Natural Categories* (Berkeley: University of California, Language Behavior Research Laboratory, Working Paper no. 43, 1975).

Color perception and categories. Trevor Lamb and Janine Bourriau, eds., *Colour: Art & Science* (New York: Cambridge University Press, 1995); Philip E. Ross, "Draining the language out of color," *Scientific American* pp. 46–47 (April 2004); Terry Regier, Paul Kay, and Naveen Khetarpal, "Color naming reflects optimal partitions of color space," *Proceedings of the National Academy of Sciences, U.S.A.* 104(4): 1436–1441 (2007); A. Franklin et al., "Lateralization of categorical perception of color changes with color term acquisition," *Proceedings of the National Academy of Sciences, U.S.A.* 105(47): 18221–18225 (2008).

Later research on color perception. Paul Kay and Terry Regier, "Language, thought and color: Recent developments," *Trends in Cognitive Sciences* 10: 53–54 (2006).

Language and color perception. Wai Ting Siok et al., "Language regions of brain are operative in color perception," *Proceedings of the National Academy of Sciences, U.S.A.* 106(20): 8140–8145 (2009).

Evolution of color perception. André A. Fernandez and Molly R. Morris, "Sexual selection and trichromatic color vision in primates: Statistical support for the preexisting-bias hypothesis," *American Naturalist* 170(1): 10–20 (2007).

212–224 CHAPTER 21: HOW CULTURE EVOLVED

Definition of culture. Toshisada Nishida, "Local traditions and cultural transmission," in Barbara B. Smuts et al., eds., *Primate Societies* (Chicago: University of Chicago Press, 1987), pp. 462–474; Robert Boyd and Peter J. Richerson, "Why culture is common, but cultural evolution is rare," *Proceedings of the British Academy* 88: 77–93 (1996).

The nature of animal and human cultures. Kevin N. Laland and William Hoppitt, "Do animals have culture?," *Evolutionary Anthropology* 12(3): 150–159 (2003).

Learning culture traits by chimpanzees. Andrew Whiten, Victoria Horner, and Frans B. M. de Waal, "Conformity to cultural norms of tool use in chimpanzees," *Nature* 437: 737–740 (2005). On imitation of a chimp's motion versus watching an artifact being manipulated by the chimp, see Michael Tomasello as quoted by Greg Miller, "Tool study supports chimp culture," *Science* 309: 1311 (2005).

Use of tools by dolphins. Michael Krützen et al., "Cultural transmission of tool use in bottlenose dolphins," *Proceedings of the National Academy of Sciences, U.S.A.* 102(25): 8939–8943 (2005).

Memory capacity of birds and baboons. Joël Fagot and Robert G. Cook, "Evidence for large long-term memory capacities in baboons and pigeons and its implications for learning and the evolution of cognition," *Proceedings of the National Academy of Sciences, U.S.A.* 103(46): 17564–17567 (2006).

The nature of working memory. Michael Baltar, "Did working memory spark creative culture?," *Science* 328: 160–163 (2010).

Genes and brain development. Gary Marcus, *The Birth of the Mind: How a Tiny Number of Genes Creates the Complexity of Human Thought* (New York: Basic Books, 2004); H. Clark Barrett, "Dispelling rumors of a gene shortage," *Science* 304: 1601–1602 (2004).

The origin of abstract thought and syntactical language. Thomas Wynn, "Hafted spears and the archaeology of mind," *Proceedings of the National Academy of Sciences, U.S.A.* 106(24): 9544–9545 (2009); Lyn Wadley, Tamaryn Hodgskiss, and Michael Grant, "Implications for complex cognition from the hafting of tools with compound adhesives in the Middle Stone Age, South Africa," *Proceedings of the National Academy of Sciences, U.S.A.* 106(24): 9590–9594 (2009).

Growth rates of Neanderthal brains. Marcia S. Ponce de León et al., "Neanderthal brain size at birth provides insights into the evolution of human life history," *Proceedings of the National Academy of Sciences, U.S.A.* 105(37): 13764–13768 (2008).

History of the Neanderthals. Thomas Wynn and Frederick L. Coolidge, "A stone-age meeting of minds," *American Scientist* 96: 44–51 (2008).

The intelligence hypothesis. Michael Tomasello et al., "Understanding and sharing intentions: The origins of cultural cognition," *Behavioral and Brain Sciences* 28(5): 675–691; commentary 691–735 (2005); Michael Tomasello, *The Cultural Origins of Human Cognition* (Cambridge, MA: Harvard University Press, 1999).

Intelligences of chimpanzees and human children. Esther Herrmann et al., "Humans have evolved specialized skills of social cognition: The cultural intelligence hypothesis," *Science* 317: 1360–1366 (2007).

The qualities of advanced social intelligence. Eörs Szathmáry and Szabolcs Számadó, "Language: a social history of words," *Nature* 456: 40–41 (2008).

225–235 CHAPTER 22: THE ORIGINS OF LANGUAGE

The argument for intentionality as the predecessor of language. Michael Tomasello et al., "Understanding and sharing intentions: The origins of cultural cognition," *Behavioral and Brain Sciences* 28(5): 675–691; commentary 691–735 (2005). See also Michael

Tomasello, *The Cultural Origins of Human Cognition* (Cambridge, MA: Harvard University Press, 1999).

Uniqueness of human language. D. Kimbrough Oller and Ulrike Griebel, eds., *Evolution of Communication Systems: A Comparative Approach* (Cambridge, MA: MIT Press, 2004).

Indirect language. Steven Pinker, Martin A. Nowak, and James J. Lee, "The logic of indirect speech," *Proceedings of the National Academy of Sciences, U.S.A.* 105(3): 833–838 (2008).

Differences among cultures in conversational turn-taking differ in the pace of conversation. Tanya Stivers et al., "Universals and cultural variation in turn-taking in conversation," *Proceedings of the National Academy of Sciences, U.S.A.* 106(26): 10587–10592 (2009).

Nonverbal vocalizations: variation across cultures. Disa A. Sauter et al., "Cross-cultural recognition of basic emotions through nonverbal emotional vocalizations," *Proceedings of the National Academy of Sciences, U.S.A.* 107(6): 2408–2412 (2010).

Chomsky on Skinner. Noam Chomsky, " 'Verbal Behavior' by B. F. Skinner (The Century Psychology Series), pp. viii, 478, New York: Appleton-Century-Crofts, Inc., 1957," *Language* 35: 26–58 (1959).

Noam Chomsky quoted on grammar. Steven Pinker, *The Language Instinct: The New Science and Mind* (New York: Penguin Books USA, 1994), p. 104.

Constraint and variation in grammar. Daniel Nettle, "Language and genes: A new perspective on the origins of human cultural diversity," *Proceedings of the National Academy of Sciences, U.S.A.* 104(26): 10755–10756 (2007).

Warm climates and acoustical efficiency. John G. Fought et al., "Sonority and climate in a world sample of languages: Findings and prospects," *Cross-Cultural Research* 38: 27–51 (2004).

Genes and pitch in language differences. Dan Dediu and D. Robert Ladd, "Linguistic tone is related to the population frequency of the adaptive haplogroups of two brain size genes, *ASPM* and *Microcephalin*," *Proceedings of the National Academy of Sciences, U.S.A.* 104(26): 10944–10949 (2007).

Newly evolved languages. Derek Bickerton, *Roots of Language* (Ann Arbor, MI: Karoma, 1981); Michael DeGraff, ed., *Language Creation and Language Change: Creolization, Diachrony, and Development* (Cambridge, MA: MIT Press, 1999).

The Al-Sayyid Bedouin sign language. Wendy Sandler et al., "The emergence of grammar: Systemic structure in a new language," *Proceedings of the National Academy of Sciences, U.S.A.* 102(7): 2661–2665 (2005).

The natural order of nonverbal representation. Susan Goldin-Meadow et al., "The natural order of events: How speakers of different languages represent events nonverbally," *Proceedings of the National Academy of Sciences, U.S.A.* 105(27): 9163–9168 (2008).

The absence of a language module. Nick Chater, Florencia Reali, and Morten H. Christiansen, "Restrictions on biological adaptation in language evolution," *Proceedings of the National Academy of Sciences, U.S.A.* 106(4): 1015–1020 (2009).

236–240 CHAPTER 23: THE EVOLUTION OF CULTURAL VARIATION

Bet-hedging and the evolution of plasticity. Vincent A. A. Jansen and Michael P. H. Stumpf, "Making sense of evolution in an uncertain world," *Science* 309: 2005–2007 (2005).

Coding and regulatory genes in development. Rudolf A. Raff and Thomas C. Kaufman, *Embryos, Genes, and Evolution: The Developmental-Genetic Basis of Evolutionary Change* (New York: Macmillan, 1983; reprint, Bloomington: Indiana University Press, 1991); David A. Garfield and Gregory A. Wray, "The evolution of gene regulatory interactions," *BioScience* 60: 15–23 (2010).

Developmental plasticity and longevity in castes of ants. Edward O. Wilson, *The Insect Societies* (Cambridge, MA: Harvard University Press, 1971); Bert Hölldobler and Edward O. Wilson, *The Superorganism: The Beauty, Elegance, and Strangeness of Insect Societies* (New York: W. W. Norton, 2009).

241–254 CHAPTER 24: THE ORIGINS OF MORALITY AND HONOR

The biological foundation of the Golden Rule. Donald W. Pfaff, *The Neuroscience of Fair Play: Why We (Usually) Follow the Golden Rule* (New York: Dana Press, 2007).

The puzzle of cooperative behavior. Ernst Fehr and Simon Gächter, "Altruistic punishment in humans," *Nature* 415: 137–140 (2002).

Group selection and the evolutionary puzzle of cooperation. Robert Boyd, "The puzzle of human sociality," *Science* 314: 1555–1556 (2006); Martin Nowak, Corina Tarnita, and Edward O. Wilson, "The evolution of eusociality," *Nature* 466: 1059–1062 (2010).

Indirect reciprocity. Martin A. Nowak and Karl Sigmund, "Evolution of indirect reciprocity," *Nature* 437: 1291–1298 (2005); Gretchen Vogel, "The evolution of the Golden Rule," *Science* 303: 1128–1131 (2004).

The complex roles of humor. Matthew Gervais and David Sloan Wilson, "The evolution and functions of laughter and humor: A synthetic approach," *Quarterly Review of Biology* 80: 395–430 (2005).

Genuine altruism in humans. Robert Boyd, "The puzzle of human sociality," *Science* 314: 1555–1556 (2006).

Group selection and altruism. Samuel Bowles, "Group competition, reproductive leveling, and the evolution of human altruism," *Science* 314: 1569–1572 (2006).

Income differential and quality of life. Michael Sargent, "Why inequality is fatal," *Nature* 458: 1109–1110 (2009); Richard G. Wilkinson and Kate Pickett, *The Spirit Level: Why More Equal Societies Almost Always Do Better* (New York: Allen Lane, 2009).

Altruistic punishment. Robert Boyd et al., "The evolution of altruistic punishment," *Proceedings of the National Academy of Sciences, U.S.A.* 100(6): 3531–3535 (2003); Dominique J.-F. de Quervain et al., "The neural basis of altruistic punishment," *Science* 305: 1254–1258 (2004); Christoph Hauert et al., "Via freedom to coercion: The emergence of costly punishment," *Science* 316: 1905–1907 (2007); Benedikt Herrmann, Christian Thöni, and Simon Gächter, "Antisocial punishment across societies," *Science* 319: 1362–1367 (2008); Louis Putterman, "Cooperation and punishment," *Science* 328: 578–579 (2010).

255–267 CHAPTER 25: THE ORIGINS OF RELIGION

Belief by scientists in God. Gregory W. Graffin and William B. Provine, "Evolution, religion, and free will," *American Scientist* 95(4): 294–297 (2007).

Religion in the United States and Europe. Phil Zuckerman, "Secularization: Europe—Yes, United States—No," *Skeptical Inquirer* 28(2): 49–52 (March/April 2004).

On deism and the ultimate creation. Thomas Dixon, "The shifting ground between the carbon and the Christian," *Times Literary Supplement*, pp. 3–4 (22 and 29 December 2006).

Universal ethics and the moral law. Paul R. Ehrlich, "Intervening in evolution: Ethics and actions," *Proceedings of the National Academy of Sciences, U.S.A.* 98(10): 5477–5480 (2001); Robert Pollack, "DNA, evolution, and the moral law," *Science* 313: 1890–1891 (2006).

Cognitive predispositions to religious belief. Pascal Boyer, "Religion: Bound to believe?," *Nature* 455: 1038–1039 (2008).

Brain activity and imaginings. J. Allan Cheyne and Bruce Bower, "Night of the crusher," *Time*, pp. 27–29 (19 July 2005). A full coverage of brain function and belief in the supernatural, including religious founders and prophets, is provided by the multiple authors of *Neurotheology: Brain, Science, Spirituality, Religious Experience*, ed. Rhawn Joseph (San Jose, CA: University of California Press, 2002).

Ayahuasca dreams. Frank Echenhofer, "Ayahuasca shamanic visions: Integrating neuroscience, psychotherapy, and spiritual perspectives," in Barbara Maria Stafford, ed., *A Field Guide to a New Meta-Field: Bridging the Humanities-Neurosciences Divide* (Chicago: University of Chicago Press, 2011). The dreams cited by Echenhofer were originally recorded by the anthropologist Milciades Chaves and psychiatrist Claudio Naranjo.

Hallucinogenic drugs and religious prophets. Richard C. Schultes, Albert Hoffmann, and Christian Rätsch, *Plants of the Gods: Their Sacred, Healing, and Hallucinogenic Powers*, rev. ed. (Rochester, VT: Healing Arts Press, 1998).

The evolutionary steps to modern religion. Robert Wright, *The Evolution of God* (New York: Little, Brown, 2009).

268–284 CHAPTER 26: THE ORIGINS OF THE CREATIVE ARTS

Optimal arousal in visual design. Gerda Smets, *Aesthetic Judgment and Arousal: An Experimental Contribution to Psycho-Aesthetics* (Leuven, Belgium: Leuven University Press, 1973).

Biophilia and the preferred human habitat. Gordon H. Orians, "Habitat selection: General theory and applications to human behavior," in Joan S. Lockard, ed., *The Evolution of Human Social Behavior* (New York: Elsevier, 1980), pp. 49–66; Edward O. Wilson, *Biophilia* (Cambridge, MA: Harvard University Press, 1984); Stephen R. Kellert and Edward O. Wilson, eds., *The Biophilia Hypothesis* (Washington, DC: Island Press, 1993); Stephen R. Kellert, Judith H. Heerwagen, and Martin L. Mador, eds., *Biophilic Design: The Theory, Science, and Practice of Bringing Buildings to Life* (Hoboken, NJ: Wiley, 2008); Timothy Beatley, *Biophilic Cities: Integrating Nature into Urban Design and Planning* (Washington, DC: Island Press, 2011).

On fiction as truth. E. L. Doctorow, "Notes on the history of fiction," *Atlantic Monthly* Fiction Issue, pp. 88–92 (August 2006).

The dawn of the creative arts. Michael Balter, "On the origin of art and symbolism," *Science* 323: 709–711 (2009); Elizabeth Culotta, "On the origin of religion," *Science* 326: 784–787 (2009).

The meaning of Paleolithic cave art. R. Dale Guthrie, *The Nature of Paleolithic Art* (Chicago: University of Chicago Press, 2005); William H. McNeill, "Secrets of the cave paintings," *New York Review of Books*, pp. 20–23 (19 October 2006); Michael Balter, "Going deeper into the Grotte Chauvet," *Science* 321: 904–905 (2008).

Paleolithic musical instruments. Lois Wingerson, "Rock music: Remixing the sounds of the Stone Age," *Archaeology*, pp. 46–50 (September/October 2008).

Songs and dances of hunters and gatherers. Cecil Maurice Bowra, *Primitive Song* (London: Weidenfeld & Nicolson, 1962); Richard B. Lee and Richard Heywood Daly, eds., *The Cambridge Encyclopedia of Hunters and Gatherers* (New York: Cambridge University Press, 1999).

The relation of language and music. Aniruddh D. Patel, "Music as a transformative technology of the mind," in Aniruddh D. Patel, *Music, Language, and the Brain* (Oxford: University of Oxford Press, 2008).

287–299 CHAPTER 27: A NEW ENLIGHTENMENT

Controversy over inclusive-fitness theory. Martin A. Nowak, Corina E. Tarnita, and Edward O. Wilson, "The evolution of eusociality," *Nature* 466: 1059–1062 (2010); response by critics in *Nature*, March 2011, online.

Globalization and the broadening of personal group identification. Nancy R. Buchan et al., "Globalization and human cooperation," *Proceedings of the National Academy of Sciences, U.S.A.* 106(11): 4138–4142 (2009).

Index

Africa, breakout, 77–84
African wild dogs, 32, 137
agriculture, origins, 92–93, 101–2
Al-Sayyid Bedouin, 234
altruism, 109–30, 133–57
  see also eusociality
Ammophila (wasps), 152
amygdala, 60–61, 100
ants, 110–30, 154
aphids, eusocial, 137, 149–50
Ardipithecus, 26, 46
art, 269–82
Athenians, 65
Australians, 84
Australopithecinae, 30–35, 37, 49, 77–78

bees:
  eusocial, 136, 150–51, 155
  language, 229
beetles, eusocial, 137, 149
behaviorism, 158–59
bias, origin, 50, 137
biomass, humans and ants, 119
biophilia, 271–73
birds:
  instinct, 158–60
  intelligence, 37–38, 215
  memory, 215
  social behavior, 31–32
blank-slate theory, 217
bonobos, 40–41
Book of Revelation, 262–63
breakout from Africa, 77–84
Buddhism, 68–69
bushmen, African, 30

campsites, 31, 47–48, 226
carnivores, brain size, 39
causation, evolution, 164–65
cave art, 26, 278–81
child development, 60
chimpanzees:
  bipedal, 24–25
  culture, 213–14, 222
  fighting, 72–73
  genetics, 87
  hunting, 40–41
  intelligence, 37–39, 227–28
  social behavior, 38, 73
  war, 73–74
Chinese, incest avoidance, 201–2
civilization, origin and evolution, 97–105
classification, Old World primates (including humans), 42–44
color vocabulary, 205–10
communication, genetic variation, 90
  see also language, origins
conscious mind, 9–10
cooperation, evolution, 47–48
creation myths, 7–8, 260
creative arts, 270
Crusades, 64

culture:
  definition, 213
  origin and evolution, 13, 85–105, 192–224, 236–40

density dependence, population control, 74–76
dinosaurs, 137–38
dogs, African wild dogs, social behavior, 32, 137
dolphins, culture, 214
dominance, ecological, 110–57
dreams, origin of religion, 260–66
drugs, in religious visions, 260–66

Enlightenment:
  old, 275
  new, 287–97
environment, in human evolution, 29, 33, 37–39
epigenetic rules, 193–210, 239–40
eusociality, 17–20, 31, 40–41, 49–56, 109–19, 133–57
  new theory, 183–88
evolution:
  current and future, 94–96
  forces, 49–56, 166–88
  general principles, 158–88
  hominins, 43–44
  maze, 21–22
  new theory, 183–88

fire, in human evolution, 29–31, 47–48
Flores man, *see Homo floresiensis*
Fukomys (mole rat), 42

Gauguin, Paul, 1–5
gene-culture coevolution, 195–210
genetic diversity, human, 80–81
genetic drift, 87
genetic evolution:
  general principles, 158–65
  humans, 86–91, 94–96
  insects, 156–57
genocide, 63–67
Giliadites, 60
Göbekli Tepe, 103
grammar, 231–35
group formation, 57–61
group selection, 142–47, 156–57, 162–65, 170–88

hand, evolution, 23
haplodiploid hypothesis, 169–70
Heterocephalus (mole rats), 41–42, 137
Homo erectus, 39–41, 48, 79–80, 225–26
Homo floresiensis, 15, 78
Homo habilis, 35–37, 39, 225–26
Homo heidelbergensis, 276
Homo neanderthalensis, 15, 85–86, 217, 222–23
Homo sapiens:
  breakout and spread, 77–84
  diagnosis, 79–80
  preadaptations in evolution, 22–32
  running, 27–28
  senses, 268–69
  as a species, 15–17, 35
  see also human nature
homosexuality, 253–54
honor, 251–52
humanae vitae, 192, 252
humanities, 274–75, 287
human nature, 158–59, 191–210, 290
hunter-gathers, 30, 80
hunting, 29
incest avoidance, 199
inclusive fitness, theory, 143–47, 166–82
insects, 17–20, 49, 110–30, 138–57
  Paleozoic, 133–36

instinct, 13, 158–65
intelligence, 37–39, 217–20
intentionality, 226–29
introspection, 8–9
Islam, 64
Israeli kibbutzim, 202

kin selection, 143–47, 166–82

lactose tolerance, 198
language, origins, 225–35
learning, 213–24
literature, 275–77
locomotion, 23–29
Luther, Martin, on God and war, 64

mammals, evolution, 17–18
manufacture, 217–20
Mandan Indians, 264–65
Mayans, war, 66
meat, consumption, 30–31, 39–41, 47–48
Melos, 65
memory, 213–24
mind, 9–10, 217–24
mole rats, eusocial 41–42, 137
moral reasoning, 241–54, 289
  genetic diversity, 81
mortality, warfare, 70–71
multilevel natural selection, 49–56, 162–63, 166–88, 274, 289
music, 281–84
mutations, 88, 159–61

naked mole rats, 41–42, 137
Neanderthals, *see Homo neanderthalensis*
Neolithic innovations, 67
nests, key to eusociality, 140–42, 148–57, 225–26
New Guinea, color vocabulary, 208
New World, invasion, 83–84

Paleolithic period:
  art, 69
  genocide, 67–69
personality, variation, 100–101
phenotype/genotype, 158–65
philosophy, 1, 9–10, 191
phylogeny, hominins, 43–44
plasticity, phenotype, 163–64
preadaptations, to eusociality, 21–32, 45–48, 140–42, 148–57, 225–26
prejudice, origin, 59
primates, evolution, 43–44, 200
progressive brood care, 140–42, 148–57
protected nests, key to eusociality, 140–42, 148–57
proximate causation, 164–65

rarity, of eusociality, 133–38
reflexes, 194
religion, 1, 7–8, 191, 255–67, 291–95
Richard I, King of England, 64
running, 27–29
Rwanda, genocide, 63

science, role, 10
selfish gene, *see* kin selection
shrimp, eusocial, 137
Simon's principle, order and hierarchy, 99
social insects, 14–15, 17–20, 109–57, 238–39
  see also eusociality
sociobiology, discipline, 169
Solomon, ant wisdom, 130
space travel, 295–97
Sphecomyrma, 122
superorganism, 147

symbiosis, social, 123–30
Synalpheus, eusocial shrimp, 137, 150

termites, 120–21, 136
territory, 75–76, 114
theology, *see* religion
thrips, eusocial, 137, 149–50
Thucydides, 65
tools:
  chimpanzee, 37
  human, 39, 217–20
tribalism, 57–61

ultimate causation, 164–65

vision, evolution, 23–24
visions, religious, 260–66

war, 62–76
wasps, 149, 132, 154
wolves:
  social behavior, 32
  predation, 74–75

Yanomamo, 67, 212

About the Author

Edward O. Wilson was born in Birmingham, Alabama, in 1929 and was drawn to the natural environment from a young age. After studying evolutionary biology at the University of Alabama, he has spent his career focused on scientific research and teaching, including forty-one years on the faculty of Harvard University. His twenty-six books and more than four hundred mostly technical articles have won him over one hundred awards in science and letters, including two Pulitzer Prizes, for *On Human Nature* (1979) and, with Bert Hölldobler, *The Ants* (1991); the U.S. National Medal of Science; the Crafoord Prize, given by the Royal Swedish Academy of Sciences for fields not covered by the Nobel Prize; Japan's International Prize for Biology; the Presidential Medal and Nonino Prize of Italy; and the Franklin Medal of the American Philosophical Society. For his contributions to conservation biology, he has received the Audubon Medal of the National Audubon Society and the Gold Medal of the Worldwide Fund for Nature. Much of his personal and professional life is chronicled in the memoir *Naturalist*, which won the *Los Angeles Times* Book Award in Science in 1995. More recently, Wilson has ventured into fiction, the result being *Anthill*, a *New York Times* bestselling novel published in 2010. Active in field research, writing, and conservation work, Wilson lives with his wife, Irene, in Lexington, Massachusetts.